Understanding Optronics

By: Larry B. Masten, Ph.D.
Professor & Chairman
Engineering Technology Dept.
Texas Tech University
Staff Consultant, Texas Instruments Learning Center

Billy R. Masten, B.S.
Engineering Manager
Optoelectronics Dept.
Texas Instruments Incorporated

Managing Editor: Gerald Luecke
Mgr. Technical Products Development
Texas Instruments Information Publishing Center

Editor: Charles W. Battle
Texas Instruments Information Publishing Center

P.O. BOX 225474, MS-8218 • DALLAS, TEXAS 75265

This book was developed by:
The Staff of the Texas Instruments Information Publishing Center
P.O. Box 225474, MS-8218
Dallas, Texas 75265

For marketing and distribution inquire to:
Orm Henning
Marketing Manager
P.O. Box 225474, MS-8218
Dallas, Texas 75265

With contributions by:
Gerald Luecke
Douglas Brandon

Appreciation is expressed to Deanne Williams for manuscript typing and Yondell Masten for proofreading.

Word Processing:
Barbara Morgan
Vicki Seale

Artwork and layout by:
Plunk & Associates

ISBN 0-89512-049-6
Library of Congress Catalog Number: 81-51950

Third Printing

IMPORTANT NOTICE REGARDING BOOK MATERIALS

Texas Instruments makes no warranty, either expressed or implied, including but not limited to any implied warranties of merchantability and fitness for particular purpose, regarding these book materials and makes such materials available solely on an "as-is" basis.

In no event shall Texas Instruments be liable to anyone for special, collateral, incidental, or consequential damages in connection with or arising out of the purchase or use of these book materials and the sole and exclusive liability to Texas Instruments, regardless of the form of action, shall not exceed the purchase price of this book. Moreover, Texas Instruments shall not be liable for any claim of any kind whatsoever against the user of these book materials by any other party.

Copyright © 1981
Texas Instruments Incorporated, All Rights Reserved

For conditions, permissions and other rights under this copyright, write to Texas Instruments, P.O. Box 225474, MS-8218, Dallas, Texas 75265.

Table of Contents

Chapter		Page
	Preface	iv
1	*Light Radiation*	1-1
	Quiz	1-20
2	*Light Radiation Sources*	2-1
	Quiz	2-23,24
3	*Light Detectors*	3-1
	Quiz	3-22
4	*Optically Coupled Electronic Systems*	4-1
	Quiz	4-24
5	*Optoelectronic Displays*	5-1
	Quiz	5-33,34
6	*Applications of Light-Emitting Diodes*	6-1
	Quiz	6-22
7	*Applications of Photodetectors*	7-1
	Quiz	7-29,30
8	*Applications of Photocoupled Data Acquisition Systems*	8-1
	Quiz	8-27,28
9	*Applications of Photocoupled Data Transmission Systems*	9-1
	Quiz	9-26
10	*Applications of Lasers*	10-1
	Quiz	10-32
	Glossary	G-1
	Index	I-1

Preface

Where would we be without light? Recall the last time a thunderstorm caused a power black-out. You seemed so alone and isolated until the candle was lit. Then the world surrounded you again when the light came on.

By understanding light and its characteristics and by combining it with electronics, many useful tasks are being accomplished. If you are interested in how this is done, the devices that are used to do it, what characteristics are required of the devices, what the terms and definitions are, and the thought processes that go into applying the device to systems that solve problems, then this book should be of interest to you.

It begins with a discussion of light radiation, where it comes from, how it is detected, and the terms and definitions for describing it. Next, sources and detectors are examined to understand the types, how they work, their characteristics, and where they might be applied. With this background, the parts of total systems are discussed and the system characteristics defined. Interruptible, non-interruptible, transmission medium, transfer characteristics, frequency response, switching times, isolation, and noise are key words.

People need devices that communicate information in visual form. These are displays. The book discussion covers the types and their operations; from visible light-emitting diodes (VLEDs) to plasma to cathode-ray tubes.

Following displays, the concentration is on applications. Actual system examples are given to demonstrate how to decide what system characteristics are required, how devices are selected, and actual circuit solutions. Many times, further theory of operation is injected into the discussion to give more detail; as in the chapter on the application of lasers.

This book, like the others in the series, is designed to build understanding step-by-step. Try to master each chapter before going on to the next one, especially the first five chapters. A quiz is provided at the end of each chapter for personal evaluation of progress. Answers are included.

A glossary and index are provided to aid in using and understanding the material and finding the subjects of interest.

The physics and properties of light have been known and understood for many, many years, but the combination of light with modern-day electronics — semiconductors, integrated circuits, infrared and laser technology — creates a most exciting and expanding field. We hope this book contributes to your understanding and appreciation of it.

L.M.
B.M.
G.L.

1 LIGHT RADIATION

Light Radiation

IN THE BEGINNING

"In the beginning God created the heaven and the earth. And the earth was without form, and void; and darkness was upon the face of the deep. And the Spirit of God moved upon the face of the waters. And God said, Let there be light: and there was light. And God saw the light, that it was good: and God divided the light from the darkness."[1]

Since time began, man has been able to see light and its effect with his eyes and sense its effect with his body. Because of light, even a newborn baby senses the presence of bright and dull objects, the light of day and the darkness of night, and the brightness and heat of the sun.

As time passes, light helps a baby to learn about its surroundings—to perceive color, shape, contrast, and intensity. Light brings messages from the faces of people that the baby learns to interpret as happiness, gentleness, approval, love, hate, sadness, danger, and anger. In these cases, light is being used as a source and the reflections from objects in the surrounding environment bring data to the baby through the eyes. In other cases, the presence of light itself may bring a direct message. For example, the sunrise signals the start of a new day; the sunset signals the end of a day. Both the reflected light images and the presence of light represent data acquisition systems, that is, the ways and means of obtaining information.

Figure 1-1 illustrates how important light, particularly the amount of light, is to obtaining information. A series of pictures of the same information taken at different light levels is shown. Without a source of light, information is available but is not transferred as in *Figure 1-1a*. With a weak light source, the same information is available to be transferred, but is barely distinguishable as in *Figure 1-1b*. With an adequate light source, the information is transferred to the reader in a clear manner as in *Figure 1-1c*.

---

[1] *The Holy Bible*, Genesis 1:1-4 KJV

LIGHT RADIATION 1

This sequence depicts the effects of light intensity on the transfer of information from the paper to the viewer.

Figure 1-1. Effects of Light on Information Transfer

As a result of understanding and applying the principles of light, many useful and unusual tasks can be performed. For example, the light technology developed to guide aircraft or missiles has contributed to navigating vehicles such as the space shuttle shown in *Figure 1-2a*. This light technology and additional communications technology have permitted close range detection of unusual phenomenon such as the rings around the planet Saturn shown in *Figure 1-2b* and transmission of the information back to earth. In *Figure 1-2c*, light is used as a carrier of information and many different channels can be transmitted on a single fiber optics cable to increase data handling and data communications capability. Photographic techniques (which, of course, are based on using light) help to produce the microscopic sized large-scale integrated circuits. Their small size is apparent when compared to the size of an ant, as shown in *Figure 1-2d*. Such "solid-state" circuits have reduced the physical size of computer systems of a given computing capability by 1,000 to 5,000 times. What once required a room can now be held in the hand. Delicate surgery using laser light technology is now performed in minute places where it was never possible before.

Infrared control systems, video systems that extend the TV screen into industrial applications, solar energy collectors for heating, point-of-sale product code detectors, intrusion alarm detectors, smoke detectors, and many similar applications are further examples of the use of the characteristics of light to do useful things.

These applications may be divided into two broad categories. One is the control and use of light energy; the other is the transmission and processing of information carried by patterns of light that depend on the presence or absence of light. To be able to use light techniques effectively, a designer of systems must understand and control one or all of the following: the characteristics of the light source, the characteristics of the transmission medium (the atmosphere or material through which the light travels), and the characteristics of the detector or sensor of the light.

1 LIGHT RADIATION

a. Space Shuttle
(Courtesy of NASA)

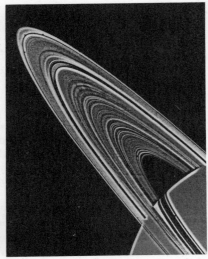

b. Rings of Saturn
(Courtesy of NASA)

c. Fiber optic Cable

d. Microphotography
(Courtesy of George Dillman, Engineering Services,
Texas Tech University)

Figure 1-2. Applications of the Characteristics of Light

LIGHT RADIATION 1

WHAT IS LIGHT?

Light is energy. It is a type of energy called *radiant energy* that travels in the form of electromagnetic waves. Radiant energy travels through space in the form of waves much like the waves produced in still water if we toss a pebble into it. The waves radiate in every direction away from the point of origination, that is, the point where the pebble entered the water. Infrared, ultraviolet, and visible light are types of radiant energy classified as light. Radio waves and X-rays are similar types of radiant energy but are not classified as light.

Radiant energy is transferred from one location (the radiator) to another (the receiver) without physical contact. For example, when you stand in front of an operating electric heater, you immediately feel the radiant heat that comes to you directly without heating the air between you and the heater. Solar energy is radiated from the sun and absorbed or reflected from objects that it strikes. When absorbed, it heats the objects that it strikes. When collected as shown in *Figure 1-3*, it can be used to heat water to produce steam and generate electric power with turbines. Microwaves are radiated in a microwave oven to cook food without heating the air. In each of these cases, no physical contact is made between the radiant energy source and the receiver.

First solar collector to generate commercial electrical power. Crosbyton, Texas.

***Figure 1-3.** Solar Energy Collector*
(Courtesy of George Dillman, Engineering Services, Texas Tech University)

1   Light Radiation

Radiation describes the energy transfer (the flow of energy) from a radiator (transmitter or emitter) to a receiver (detector or sensor). We can't see the energy transfer, but we can see its effect; just as we can't see the wind, but we can see its effect. Radiant energy transferred in this way is said to be in the invisible part of the electromagnetic spectrum. The electromagnetic spectrum contains the various radiant energy sources separated according to frequency. The infrared and ultraviolet light energy are included in the invisible part of the spectrum.

There are also radiations of light energy that are visible. Incandescent lamps, fluorescent lamps, the sun, candles, kerosene lamps, neon signs, and electronic displays are examples of light energy radiation in the visible spectrum.

The primary difference between all of these types of radiant energy is their frequency, as shown in the graphic display of the electromagnetic spectrum in *Figure 1-4*. (We will tell you what the terms mean a little later.) As you can see, visible light is only a small portion of the spectrum of radiant energy. In this book, the term "light" includes sources that are at the infrared and ultraviolet frequencies as well as visible light rays.

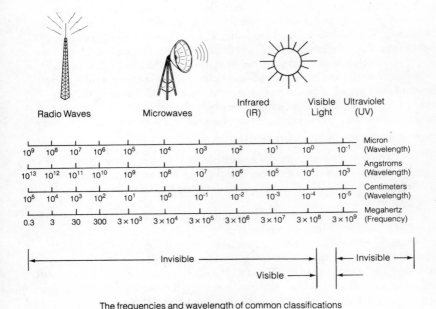

The frequencies and wavelength of common classifications of radiant energy are compared.

***Figure 1-4.** The Electromagnetic Frequency Spectrum*

LIGHT RADIATION 1

WHERE DOES LIGHT COME FROM?

Light is produced by the release of energy from the atoms of a material when the atoms are excited by heat, chemical action, or other means. The tiny bundles of energy are called photons and light consists of a stream of these photons. For years, scientists argued whether photons of light behaved like waves or like particles being shot from a gun. Finally, they have agreed that actually light has both characteristics, but most ideas about light can be described in terms of waves.

Natural Light Sources

The sun is our most important natural source of light. Sunlight is produced by the extreme heat generated by atomic reactions (much like a nuclear furnace) in the sun. The sun produces radiant energy not only in the visible light band but also across a wide band of the electromagnetic frequency spectrum. Some would also say that moonlight is another natural light source. However, it is just sunlight reflected from the moon's surface.

Artificial Light Sources

Humans have developed many artificial light sources that make it possible to continue activities when sunlight is not present. The first was fire—the campfire and torch—which has been refined in variations such as the candle, kerosene lamp, and gas lamp. The most widely used artificial light sources are the incandescent and fluorescent electric lamps.

Another important artificial light source is the laser. This device is different from most other sources because the light is in the form of a narrow beam of photons, all of which have the same energy and wavelength. This characteristic produces a pure light of a single color with the energy highly concentrated in a narrow beam that does not spread very much as it travels.

DEFINITIONS OF UNITS AND TERMS

Cycle is a measure of the repetitiveness of a waveform. It is indicated when a waveform starting at a point on its variation returns to that same point on the waveform and continues to repeat itself in every respect.

Wavelength is a measure of the distance a wavefront traveling at the speed of light travels in space in one cycle (*Figure 1-5*). Long and medium wavelengths of radiated waves are measured in ordinary units of distance measure such as the mile (kilometer), yard (meter), and inch (centimeter); however, the wavelengths of light are very short and are normally measured with a special unit called the *angstrom* which is abbreviated Å. One angstrom is equal to one ten-billionth of a meter or about 1/250,000,000 of an inch (*Figure 1-6*).

1 LIGHT RADIATION

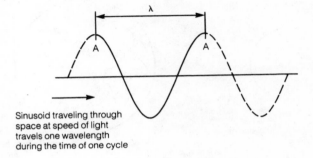

Figure 1-5. *Wavelength*

(Source: *Understanding Communications Systems*, D.L. Cannon and
G. Luecke, Texas Instruments 1980)

Period is a measure of the time it takes for one cycle to occur. The second or fractions of a second such as, millisecond (0.001 second), microsecond (0.000001 second), nanosecond (0.000000001 second), and picosecond (0.000000000001 second) are used to measure the period of radiant energy waves.

Frequency is a measure of how many cycles of the wave occur in one second. In earlier days, the unit of frequency measurement was cycles per second, but it has been renamed hertz. For large values of frequency, prefixes are used with hertz so that kilohertz means 1,000 hertz, megahertz means 1,000,000 hertz and gigahertz means 1,000,000,000 hertz.

Frequency and period are related by the equation

$$f = \frac{1}{t}$$

where f is the frequency in hertz and t is the time (period) in seconds.

LIGHT RADIATION 1

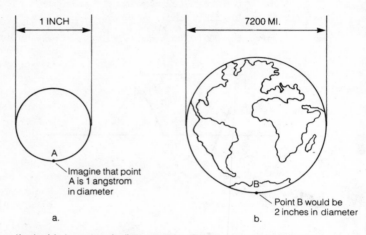

If point A is 1 angstrom in diameter and could be seen on a 1 inch diameter circle, then when the circle is expanded to the diameter of the earth, point A would be a point (point B) that would have a diameter of 2 inches.

Figure 1-6. Length of Angstrom Comparison

The *speed of light* is extremely fast and has been determined to be 186,282 miles (299,792 kilometers) *per second* in a vacuum or in free space outside the earth's atmosphere. However, when light passes through a medium (material), such as glass or water, the speed decreases slightly. This characteristic is put to good use in the design of lens for microscopes, telescopes, and ordinary eyeglasses. Frequency, wavelength, and speed are related as shown in this equation and *Figure 1-5:*

$$\text{Wavelength (meters)} = \frac{\text{Speed of light (300,000,000 meters per second)}}{\text{Frequency (Hz)}}$$

Note that the speed of light is rounded to make calculations easier unless precision is required.

The speed of light is much faster than the speed of sound because sound must have air or some other medium to support its travel while light does not. Most of us have observed the starter at a track meet and have seen the smoke from the starting gun before the sound is heard; or we've been in a thunderstorm and have seen the lightning flash before the thunder is heard. This phenomenon is observed because the speed (velocity) of sound, approximately 331.4 meters per second, is slow compared to the speed of light, 300,000,000 meters per second.

1 LIGHT RADIATION

HOW DO WE DETECT RADIANT ENERGY?

To be useful, radiant energy of all forms requires some kind of detector or sensor to detect its presence and to convert it to another form of energy. The eye is a natural visible light detector, but people have developed detectors for most frequencies of radiant energy. For example, a detector is used in home radio and television receivers to detect the radio waves that are transmitted by a local station to convert them into useful electrical energy at the receiver. Detectors have also been developed to detect the presence of infrared, visible, ultraviolet, and X-ray frequencies. The way these detectors convert radiant energy to other forms of energy (often electrical) will be discussed in later chapters.

Since the primary difference between the kinds of radiant energy is frequency and the related wavelength, then it follows that the primary difference between the various detectors is that they respond only to certain frequencies. This allows a detector to be designed that produces an output only when the selected frequencies of radiant energy are present.

As a simple illustration of this, suppose one wishes to monitor the presence of waves in water as shown in *Figure 1-7*. Two floating vessels are selected as detectors. One is a small 10-foot fishing boat, the other is an oil tanker. The presence of waves is to be detected by observing the motion

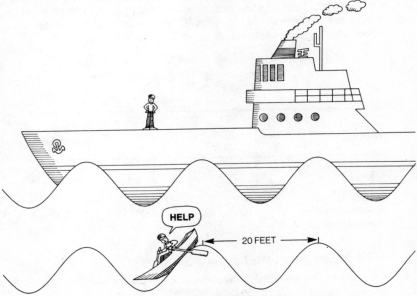

Illustrates the response of small boat (detector) to wavelength compared to the response of the large boat (also detector) to the same wavelength.

***Figure 1-7.** Frequency Sensitivity of Detectors*

UNDERSTANDING OPTRONICS

LIGHT RADIATION 1

of the boats. The water is disturbed so that a wave is produced with a wavelength of 20 feet. It is observed that the 10-foot fishing boat rises and falls with each wave (produces an output) while the oil tanker is unaffected by the waves (does not produce an output). Therefore, we observe that the two detectors (vessels) respond differently to the same wave energy. In a similar manner, different light detectors respond differently when stimulated by the same light energy.

How Do We Detect Color?

The human eye responds only to wavelengths between approximately 4,000 and 7,000 angstroms, which, as shown in *Figure 1-8* and *Figure 1-4*, is that portion of the electromagnetic frequency spectrum we call visible light. If the light seen by the eye contains approximately equal amounts of energy for all wavelengths in this range, the eye produces an output that is sensed by the brain as white light. If certain wavelengths have more energy than others, the eye's output is sensed as one or more colors. As shown in *Table 1-1*, color is a set of names given to the response of the human eye to different wavelengths (frequency) of light in the visible range. The color we call red has the longest wavelength at about 7,000 angstroms and violet has the shortest at about 4,000 angstroms. You may have seen an experiment that passes white light through a glass prism which breaks up the white light into the various colors that progress from red to orange to yellow to green to blue to violet. If not, you have observed the same effect, although not as sharply defined, in a natural rainbow.

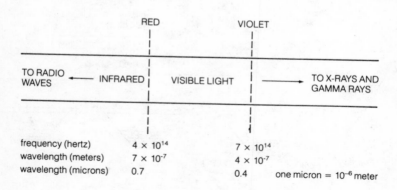

Figure 1-8. *Visible Portion of Spectrum of Figure 1-4*

(Source: Understanding Communications Systems, D.L. Cannon and G. Luecke, Texas Instruments 1980)

1 LIGHT RADIATION

Table 1-1. Relationship of Colors to Wavelength

Color	Wavelength
Violet	Below 4500Å
Blue	4500-5000Å
Green	5000-5700Å
Yellow	5700-5900Å
Orange	5900-6100Å
Red	6100-7000Å
White	Is the equal combination of all wavelengths
Black	Is the total absorption of all wavelengths

Colored light can also be produced by passing white light through a material such as colored glass or plastic. This material is called a filter because it stops or absorbs some wavelengths of the white light while allowing other wavelengths to pass through.

How Do We Detect the Presence of Objects?

So far the light we have been considering has come from a source such as the sun or electric lamp. But we can also detect light that is reflected. In fact, this is how we see things; that is, a light source illuminates or shines on objects and we see the object because of the light reflected from it. If the object reflects all wavelengths equally, the eye sees the object as having a white color (assuming a white light source). If the surface of the object absorbs some wavelengths while reflecting others, the eye senses a color as shown in *Figure 1-9*. If the object absorbs all wavelengths of the white light, it is sensed as being black in color.

This also explains why black objects get hotter than other colored objects, particularly white, when exposed to sunlight. Remember that light is a form of energy. Therefore, when the black object absorbs the energy, it is transformed into heat which heats the object. Since a white object reflects the energy it doesn't get as hot. In fact, the heating of the white object is produced mainly by absorption of energy from radiant waves outside the visible light range.

LIGHT RADIATION 1

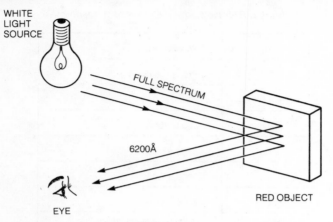

The red surface absorbs all wavelengths except the 6200Å light.
The eye responds to the reflected 6200Å light and with the help of the brain calls it red light.

Figure 1-9. Effect of Colored Object on White Light

HOW IS LIGHT INTENSITY MEASURED?

Another characteristic of light that is important is intensity. Intensity is a measure of the energy contained in radiated light waves. There are two main methods of measuring intensity; one, the photometric system, is related to the eye and the other, the radiometric system, is related to human-made detectors. The photometric system is used to measure intensity in terms of the sensation of brightness as sensed by the human eye. The radiometric system is used to measure intensity in terms of energy, that is, the ability to do work.

Photometric System

The photometric system for measuring intensity was developed to relate light measurements to the human eye because the eye is a special type of detector. Its job is to "see"; that is, to determine the position, shape and color of objects by converting an image in light to electrical impulses that are used by the brain.

A brief review of the eye may aid in the understanding of the photometric system. The lens of the eye (*Figure 1-10a*) is adjustable to allow light images to be focused on the retina at the back of the eye. The pupil of the eye is also adjustable to determine the intensity or brightness of the light reaching the retina. The aperture or hole of the pupil through which light passes is small for bright light images and large for dim light images. This aperture control allows the average intensity at the retina to be held constant over about a 16 to 1 variation in the brightness of the viewed image.

1 LIGHT RADIATION

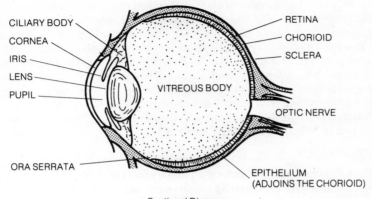

a. Sectional Diagram

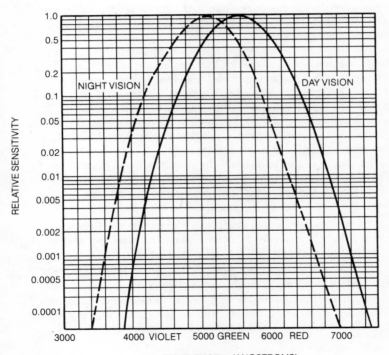

b. Sensitivity as Function of Wavelength

***Figure 1-10.** The Human Eye*

LIGHT RADIATION 1

The retina consists of about 100 million rod-shaped cells and 7 million cone-shaped cells. The rods respond to dim light (below 0.001 candela per square meter) but do not sense color. The cones, which are concentrated in the central part of the retina, require brighter light (about 10 candela per square meter) for stimulation but they provide color response and are able to distinguish between very small objects. The cones are able to sense color because they are sensitive to wavelength. The overall response of the cones varies upward from zero below 7,000 angstroms (red) to a peak at about 5,500 angstroms (green) then gradually drops to zero again above 4,000 angstroms (violet) as shown in *Figure 1-10b*.

Terms and Units

Let's define the terms and units used in the photometric system. The amount of light produced by a light source is called the *luminous intensity*. The standard unit used to measure luminous intensity is now the *candela*. (For many years, it was called the "candle" because the luminous intensity of a certain size candle made from the wax of sperm whales was used as the standard. The term "candlepower" has also been used to describe luminous intensity).

One candela is the amount of light that shines out through a hole in one side of a ceramic box after the box has been heated to 1772°C (3222°F). The hole area is 1/60 of a square centimeter and the box is wrapped in platinum. It is heated until the platinum melts, then the box is cooled until the platinum just begins to harden. The temperature at this point is 1772°C and the ceramic inside glows with intense light which shines out through the hole in the box. The candela is used to calculate the lumen and foot-candle which are other units of light measurement.

The lumen is used to measure the amount of energy in a beam of light. The foot-candle (or lux in the metric system) is used to measure illuminance; that is, the amount of light shining on a surface. The relationship between the candela, lumen, and foot-candle is:

> A 1 candela light source produces a 1 lumen beam of light which provides 1 foot-candle of illumination on a 1 square foot area which is located on a radius of 1 foot from the source.

The intensity of light falling on a surface varies inversely (oppositely) with the square of the distance between the source and the surface. This means that if a surface that receives 1 foot-candle illuminance at 1 foot is moved 2 feet away from the surface, the surface will only receive 1/4 foot-candle illuminance. This is illustrated in *Figure 1-11*.

1 LIGHT RADIATION

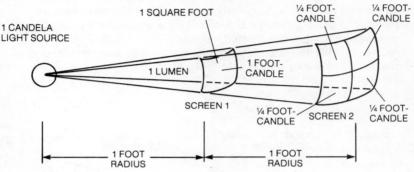

Figure 1-11. Relationship of Photometric Units

Radiometric System

The visible spectrum is a very small portion of the electromagnetic radiation spectrum. Although the human eye responds to frequencies in the visible spectrum, it does not respond to radiation in the other portions of the spectrum and cannot be used to detect their presence. As mentioned previously, human-made detectors that can detect these other portions of the spectrum have been developed. For this reason, the radiometric system is a system to measure the radiant energy over the total spectrum.

Terms and Units

For most detectors other than the eye, the most convenient measurement system is the radiometric system. Some important terms and units in this system are defined below:

  a. A *watt* is a unit of measure of the rate at which energy is radiated or used with respect to time. Detectors respond to this rate of change and convert a portion of the radiated electromagnetic energy to a useable form such as electrical, chemical or thermal.
  b. The *intensity* of the radiation at a detector is the ratio of the number of watts striking the detector to the area of the detector.
  c. The *efficiency* of the detector is the ratio of the number of watts converted to usable output by the detector to the number of watts striking the detector.

LIGHT RADIATION 1

These terms and units are illustrated in *Figure 1-12*. The 1 watt light source is assumed to project 1 watt of light uniformly onto a 1 square centimeter screen one meter away. The intensity at screen 1 is 1 watt per square centimeter. With screen 1 removed, the intensity at screen 2 is 0.25 watt per square centimeter. A detector with an area of 0.01 square centimeter placed two meters away from the source would receive $0.01 \text{cm}^2 \times 0.25 \frac{\text{watts}}{\text{cm}^2}$ or 0.0025 watt of light, while the same detector placed one meter away would receive $0.01 \text{cm}^2 \times 1 \frac{\text{watt}}{\text{cm}^2}$ or 0.01 watt of light.

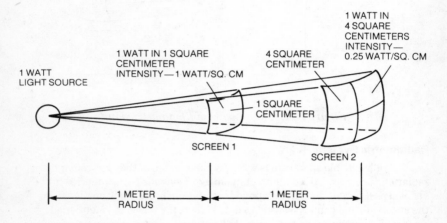

Figure 1-12. *Relationship of Radiometric Units*

HOW IS LIGHT USED?

The use of light always requires a source of light radiation, a medium through which the light travels, and a detector to convert the light to another form of energy. The processing of the light energy can be accomplished by controlling any combination of the parts of the system.

The uses of light can be divided into three categories; illumination, energy transmission and information transmission.

Illumination

Illumination requires a light source with sufficient intensity and particular wavelength characteristics so that the light reflected by an object can be detected by an appropriate detector. The detector then provides the desired information about the environment that is illuminated.

1 LIGHT RADIATION

Headlights of an automobile are used to provide information to the driver about the roadway in the immediate path of the automobile. The effectiveness of the illumination provided by the lights depends not only on the power of the headlights, but also on the road surface (black top or concrete which determines the amount of light reflected), atmospheric conditions (fog, dust or snow which affect the transmission of the light), and clearness of windshield (which affects the transmission of the light).

To properly illuminate a room for human use requires a light source with sufficient intensity and scattering to reflect light from all objects that need to be seen. The source should contain all wavelengths in the visible spectrum so that the true colors of the objects can be determined. At other times, the illumination of a room may be for special effects rather than just vision. In such cases, the intensity and color of the illumination can be used to provide a mood of softness, eerieness, daytime, nightime, etc.

To use light for machine applications, the illumination required is normally restricted to determine one or two of the following: 1) the presence or absence of an object, 2) the shape of an object, 3) the light patterns produced by an object, 4) the color of an object, or 5) the motion of an object. Each of these applications may require a different method of illumination and a different detector to obtain the desired information about the object.

Energy Transmission

In order to utilize light to transmit energy from a power source, the light source should be efficient and intense, a low-loss transmission medium should be used, and the detector should respond to the wavelengths that it receives so that it efficiently converts the light energy to the desired useful form of energy such as thermal, chemical, or electrical energy.

The sun as a source of radiant energy provides approximately 1,400 watts per square meter outside the earth's atmosphere. However, at the earth's surface the power density is reduced to approximately 800 watts per square meter due to losses in transmission through the earth's atmosphere. A solar cell array of the type shown in *Figure 1-13* has a conversion efficiency of about 10 percent; that is, it produces about 80 watts of electrical energy per square meter of solar cells from the 800 watts per square meter of solar energy. The other 90 percent of the solar energy is either reflected or is converted to heat rather than electrical power. If a black non-reflective material is used instead of a solar cell array, almost all of the 800 watts would be converted to heat. In fact, this is the method often used to convert solar energy to heat for use in homes and other buildings.

Human-made light sources are seldom used to transmit high levels of energy except in the case of some laser applications. With laser sources it is possible to concentrate the energy in such a small area that the power per square meter is high enough to scribe very hard materials, cut metals, or weld.

LIGHT RADIATION 1

Figure 1-13. *Solar Cell Array for Direct Conversion of Light to Electrical Energy*

(Courtesy of George Dillman, Engineering Services, Texas Tech University)

Information Transmission

Since both the source and the transmission medium determine the amount of light received by the detector, it is possible to convey information about both the source and the medium to a detector. In the first simple case, where the medium characteristics do not change, the source can be modulated with the desired information and the detector designed to detect the information. Since the transmission medium doesn't change, it is known that any information received is from the source. This might be referred to as a non-interruptable application and would include fiber optic links (like those shown in *Figure 1-2c*), optical isolators for machine-to-machine communication and lighted displays for machine-to-people communication. Such machine-to-people communications include warning indicators, digital clock displays, digital displays for radios, and the like.

1 LIGHT RADIATION

The other simple case would be for the source to provide illumination while the medium is modulated (interrupted). In this case the detector would be designed to detect the modulated light while suppressing effects of ambient (surrounding) light conditions. This might be referred to as an interruptible application and would include intrusion alarm systems, card readers, bar code readers (*Figure 1-14*), shaft encoders, smoke detectors, holograms, television, motion pictures, and computer graphics.

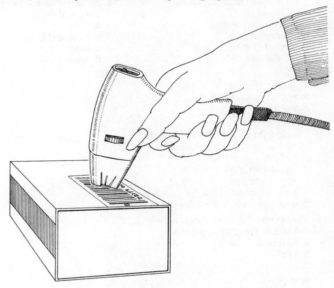

***Figure 1-14.** Bar Code Reader for Converting Bar Codes to Digital Information for Use by Computers*

WHAT HAVE WE LEARNED?

1) Light is a form of energy.
2) Light has intensity and wavelength and thus, a frequency.
3) The eye detects only a small portion of the electromagnetic spectrum.
4) There are two systems of terms and units for describing and measuring light characteristics—Photometric and Radiometric.
5) Human-made detectors are usually designed to respond to different frequency ranges of the electromagnetic spectrum.
6) Light can be used for illumination, energy transmission, and information transmission.

LIGHT RADIATION 1

Quiz for Chapter 1

1. Light is a form of
   a. work.
   b. energy.
   c. heat.
   d. force.

2. A solar water heater is an example of
   a. energy conversion.
   b. energy storage.
   c. control of energy.
   d. all of above.

3. Radiation is
   a. energy transfer.
   b. energy conversion.
   c. work.
   d. frequency.

4. Visible light is radiation of energy of _____ in the electromagnetic spectrum.
   a. all wavelengths
   b. all frequencies
   c. a narrow band of low frequencies
   d. a narrow band of high frequencies

5. The speed of light is approximately 300,000,000 meters/second in
   a. all media.
   b. vacuum.
   c. iron.
   d. water.

6. The eye is an optical
   a. source.
   b. sensor.
   c. lens.
   d. emitter.

7. Color is a result of the eye's response to
   a. wavelength.
   b. energy.
   c. distance.
   d. light.

8. The photometric system is based on
   a. response of the human eye.
   b. human-made detectors.
   c. energy density.
   d. color.

9. The radiometric system is based on
   a. response of the human eye.
   b. human-made detectors.
   c. total radiant energy.
   d. color.

10. An incandescent lamp converts electrical energy to
    a. light.
    b. light and heat.
    c. heat.
    d. work.

1. b, 2. d, 3. a, 4. d, 5. b, 6. b, 7. a, 8. a, 9. c, 10. b

2   Light Radiation Sources

Light Radiation Sources

WHAT IS A LIGHT SOURCE?

When energy reacts with any material, an energy conversion process takes place. A material which converts a portion of the energy to light is referred to as a light source. There are many sources of light in the solar system but the sun is the major source of natural light and other radiant energy for earth because it is the nearest. The sun converts atomic energy into electromagnetic radiation across a broad spectrum of wavelengths which include light rays. When this radiant energy reaches the earth's atmosphere, some wavelengths are absorbed by the earth's atmosphere and never reach the earth's surface. The energy that does reach the earth's surface in the form of infrared, visible and ultraviolet light energy is either reflected back into the atmosphere or is converted to other forms of energy. Of course, the energy from the sun which is in the visible spectrum is used directly and immediately for illumination.

The energy from the sun heats the land masses, oceans, and atmosphere. The heat causes evaporation of water from the earth into the atmosphere. When the water vapor condenses and returns to earth as rain or other precipitation, some of the water is trapped in reservoirs built by people. Since this stored water can be converted to electrical energy, the stored water is a form of stored solar energy. Energy from the sun also supports plant and animal life which stores the energy in the form of combustible products such as wood, coal, oil, and tallow. The stored energy is collected and converted over a long period of time but can be reconverted to heat and light in relatively short periods of time. *Figure 2-1* illustrates several ways by which the solar energy is stored and ultimately converted to artificial or human-made light. Therefore, people have the ability to use the stored energy for heat and light when and where they choose. Sources which can be controlled by people are referred to as human-made or specialized sources.

LIGHT RADIATION SOURCES

2

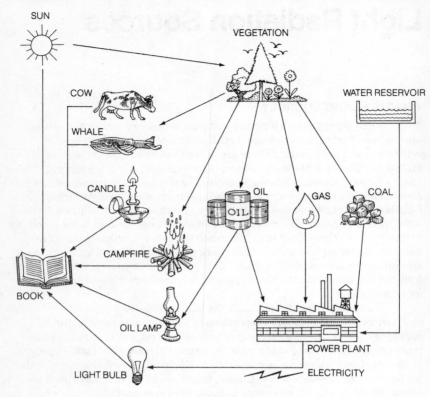

Illustration of several processes by which solar energy may be stored and reconverted to light.

***Figure 2-1.** Solar Energy Conversion to Light*

Natural Light Sources

Natural light sources are characterized by the fact that people must use the light when and where it is available. Obvious examples are the sun, moon, and stars. The sun supports all life on the earth and provides adequate illumination during the daytime. The moon provides minimal illumination at night and, as mentioned in Chapter 1, can be considered as a reflecting source. The use that people make of natural light sources depends on the intensity and wavelengths provided by the source.

2 LIGHT RADIATION SOURCES

For example, although the energy received from the stars is not sufficient to provide adequate illumination or support life, it provides information which is essential for celestial navigation. Other natural sources may also include objects which emit nonvisible radiation such as radio waves or heat waves. Heat waves, for example, may be detected and converted to visible information using infrared technologies.

Specialized Light Sources

Specialized light sources are sources that can be controlled by people and, as such, are characterized by the fact that people have adapted them for use when and where they are needed. As a result, some of the characteristics of light sources which are important in applications include light conversion efficiency, maintenance, safety, lifetime, and portability. These are compared for several different sources in *Table 2-1*. Each light source has both advantages and disadvantages for a given type of application.

Table 2-1. *Characteristics of Some Specialized Light Sources*

Light Source	Light Conversion Efficiency	Maintenance	Danger	Lifetime	Portability
Campfire	M	H	H	L	L
Torch	L-M	H	H	L	H
Candle	L	M	M	L	M
Oil Lamp	L-M	M	M	M	H
Lantern	M	M	M	M	H
Incandescent Lamp	H	L	L	M	H
Neon Lamp	H	L	L	H	L
Semiconductor Lamp	L	L	L	H	H
Radioactive Lithium, Tritium and Radium	L	L	H	H	H

L = Low; M = Medium; H = High

LIGHT RADIATION SOURCES 2

Fire

For example, let's take a campfire. When used solely for light, most of the energy from a campfire is wasted because most of the energy from the wood is converted into heat. Using a campfire for a source of light is inconvenient because the fire must be maintained, which requires a constant search for fuel, adding of fuel, and disposal of ashes. One of the wood pieces from a campfire can be used as a torch to provide portable light but it too requires constant attention and care. Open flames, such as the campfire, torch, and even candles are dangerous because of the destruction they can cause if not handled with care. So we see that open flames are not a good light source because they may be hard to light, difficult to keep lighted, difficult to extinguish or reuse, and must always be used with caution.

For use in enclosed dwellings, oil lamps with a wick to transport controlled amounts of fuel from a reservoir to the flame, and a transparent globe to provide safety and shielding for the flame from gusts of air, overcame many of the disadvantages of torches and candles. Fuel additions are not required very often and the wick requires very little maintenance. Lighting the lamp is relatively easy but does require a source of fire which is usually supplied by a match. The lamp must be placed so that it is relatively easy to reach in order to light and extinguish. Also, the lamp can be dangerous since the fuel is close to the flame.

Incandescent Lamp

The incandescent lamp (light bulb) shown in *Figure 2-2* was invented by Thomas A. Edison in the *1870's*. Because it provides artificial light, it has contributed to our present standard of living as much as any other development in modern times. It has doubled or tripled the productive hours available in a day. Education, business, and industry can produce goods and services 24 hours a day. Medical care or other emergency service is available 24 hours a day. People can work all day and, because of the availability of good artificial lighting, can enjoy travel, education, and entertainment after dark. In most large cities, the pace hardly slows at sundown. In fact, in many cases, it quickens. The availability of good light and controllable energy is often not fully appreciated until a power failure darkens our homes and cities.

2 LIGHT RADIATION SOURCES

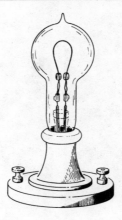

Figure 2-2. Edison's Incandescent Lamp

The incandescent lamp overcomes many of the faults of the oil lamps as light sources. The source of energy is an electrical power station which converts the stored energy of fossil fuels or water reservoirs into electricity. The energy is transmitted through a grid of electrical power lines to the customer and is ready to use at the flick of a switch. The incandescent lamp is easy to turn on and off, has fairly long life, is cheap and easy to replace, and is safe. One disadvantage of the incandescent lamp is that it produces a relatively large amount of heat which can be a fire hazard if installed improperly. Incandescent flood lamps used to light stages and color television studios generate tremendous amounts of undesirable heat.

Gas Discharge Lamps

Gas discharge lamps, such as the neon and fluorescent lamps shown in *Figure 2-3*, produce less heat for the same amount of light than the incandescent lamp but require more complex electrical circuitry. Neon lights have added a dimension of glamour and showmanship to advertising while fluorescent lights add a calm and professional atmosphere to homes, offices, and stores.

LIGHT RADIATION SOURCES 2

a. Neon

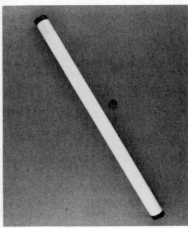
b. Fluorescent

Figure 2-3. Gas Discharge Lamps

Other light sources which have been adapted to special applications include flash bulbs which produce an intense, short-duration flash to illuminate an object for a camera. The disadvantage of the filament type flash bulb *(Figure 2-4a)* is that it is good for only one flash. A xenon flash lamp *(Figure 2-4b)*, which is another type of gas discharge lamp can be used repeatedly but requires high voltages and special circuitry to operate.

a. Filament

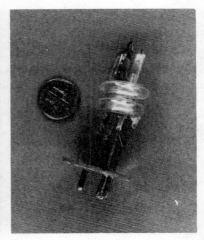
b. Xenon

Figure 2-4. Flash Bulb

2 LIGHT RADIATION SOURCES

Semiconductor Light Sources

Relatively new light sources which produce light by electricity are the semiconductor light sources which are a principal part of this book. The common part is the semiconductor diode, shown in *Figure 2-5*, called an LED for light emitting diode. Electrically it is a two element device that conducts in one direction; therefore, the name diode. It is made from semiconductor material that emits light when a certain amount of electrical current flows in the diode. An LED can be made to emit either visible or invisible (infrared) light rays.

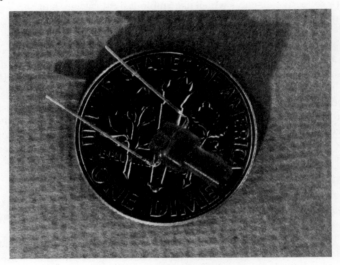

Figure 2-5. Light Emitting Diode

The visible light emitting diode (VLED) is a light source often used as an indicator lamp, either singly or in an array. An array consists of several VLEDs placed so that a pattern is presented when certain ones are turned on. VLEDs are very reliable but produce a limited color range and may be difficult to see in sunlight or other high ambient light conditions. The VLED is most often used as an equipment on-off, data, or control indicator in single units; or in calculator, clock, or watch displays in arrays. The use of VLEDs in watches has been limited because of their large power requirement compared to the power available from the necessarily small battery.

UNDERSTANDING OPTRONICS

LIGHT RADIATION SOURCES

2

Liquid Crystals

Liquid crystal elements are actually not light sources, but are light filters. Light passes through a medium or does not pass through a medium depending upon an electrical bias to the medium. Ambient light must be present for the difference to be detected. Electrodes on each side of the medium determine the areas to be energized to display numbers, or letters, or special symbols depending on the arrangement of the electrodes. The liquid crystal display (LCD) arrays are being used more in battery operated calculators and watches because of their low power consumption and because they are reflective displays which use ambient light rather than being washed out or degraded by it.

Lasers

Another special light source is the laser. Some lasers are related to semiconductor light sources because they use single crystal material. The laser produces single frequency coherent light focused in very narrow beams and has many specialized applications in communications, industrial processes, medical and surgical techniques, and research. Video disc players use lasers to read prerecorded information from a disc. Since there is no frictional contact, the disc may be played many times without the wear and degradation caused by a stylus.

Brightness Comparison

Table 2-2 compares the brightness of several common light sources. The brightness is expressed in radiometric units of milliwatts per square centimeter and in photometric units of candela per square meter.

This comparison also points out a significant difference between the photometric and radiometric units. Notice there is not a constant factor which relates the milliwatts per square centimeter to candela per square meter. From Chapter 1, recall that the radiometric units include the response to power at all radiant wavelengths while the photometric units include only the power in the visible wavelengths. For example, comparing the radiometric column, the sun produces approximately 4,500 times more milliwatts per square centimeter than a candle flame. However, looking at the photometric column, the sun produces a sensation of brightness almost 150,000 times more than the candle when the human eye is used to make the comparison. This means that sunlight has a larger percentage of power in the visible wavelengths than the candle.

2 LIGHT RADIATION SOURCES

Table 2-2. Comparison of Light Source Brightness

Source	Milliwatts/cm²	Candela/m²
Candle flame	1×10^4	1.03×10^4
Sun	4.5×10^6	1.5×10^9
Tungsten lamp		
miniature	6×10^4	2.1×10^6
standard	1.3×10^5	8.6×10^6
standard	2.6×10^3	1.7×10^5
photo	2.5×10^5	3.1×10^7
Neon	8×10^{-2}	1.6×10^1
Fluorescent	3×10^1	6.9×10^3
Xenon flash	1.2×10^7	3.4×10^9

HOW DO LIGHT SOURCES WORK?

Picture a calm pool of water on a still day. A stone is thrown into the pool. As it enters the water, the stone displaces water which causes a wave to be emitted from the point of entry and propagated or spread across the pool. Notice that the motion of the water where the stone activated it caused the wave. Light, which consists of electromagnetic waves, is emitted in a similar manner when an electron in an atom is caused to vibrate for any reason.

Think again of the stone being thrown into the water. A wave front is produced which moves from the point of impact because of the energy transferred from the stone to the water. If no more stones are thrown in the water, the wave will pass the observer and the water will again be calm. Similarly when an electron in an atom is caused to vibrate at the frequency of visible light by the addition of a pulse of energy, a light wave is generated and propagates from the electron. A detector (observer) detects a short burst of light.

On the other hand, if the stone were tied to a string and vibrated in and out of the water in a regular pattern, then the energy in the form of water waves would continue to be transmitted. Similarly, if the electron in an atomic structure is caused to vibrate continuously by continuous addition of energy, then the detector detects a continuous source of light.

Therefore, any mechanism that causes an electron to vibrate in an atom will cause the emission of a stream of electromagnetic waves. If it is vibrated fast enough so that the wavelengths are in the range of the electromagnetic spectrum called light, then light is emitted. Virtually any mechanism which results in energy being added to an appropriate substance can cause light emission.

LIGHT RADIATION SOURCES

2

Mechanisms to Produce Light

The most common and earliest used mechanism to produce light was heat. The incandescent bulb, hot coals, fire, and gas lamps are examples of sources which use this activation technique. A substance is heated to a temperature sufficiently high so that the electrons in the atomic structure of the atoms are excited to higher energy orbits and decay to lower energy orbits. Since each element has only a limited number of distinct orbits that it will allow the electron to occupy, light emitted from a particular atom will have only those wavelengths associated with that atom. This fact is often used to identify different compounds. For example, copper produces a green light.

Recall that the atom is composed of electrons which orbit a nucleus. The number of electrons and protons determine the type of material as well as the electrical and chemical properties of that material. The orbits of the electrons are also called shells and these shells are arranged as shown in *Figure 2-6a*. Each shell has a maximum number of electrons that it can hold. The K shell holds a maximum of two electrons, the L shell holds a maximum of eight electrons. The maximum number of electrons in the M and N shell is eighteen and thirty-two respectively. In addition to these maximum numbers, another necessary rule is that the outer shell of any atom has a maximum of eight electrons. In fact, the periodic table of elements used in chemistry and physics uses the number of electrons in this outer shell to define the groups of elements. This outer shell is called the valence shell.

Also recall that if this valence shell contains only one electron, the material is a good conductor; that is, very small amounts of external energy are required to remove the electron from the outer shell and make it a "free" electron which can move in the molecular structure of the material. Copper, gold, and silver have only one valence electron and are good conductors. The other important property of the valence shell is that if it is full (eight electrons), a large amount of external energy is required to pull an electron from the shell. Iron, cobalt, and platinum have eight valence electrons and are poor conductors. Other materials such as silicon and germanium have four valence electrons (*Figure 2-6b*) and are called semiconductors.

The forces in the atom are very complex because of the forces of attraction between the nucleus and electrons; the forces of repulsion between the electrons in the various shells, and the forces, due to the rapid motion of the electrons in their shells. The forces also depend on the neighboring atoms. As a result of these forces, there are discrete energy levels which can be associated with each shell. If the atomic structure is disturbed by adding energy in any form (heat, light, or electric field), the electrons may gain enough energy to move to a higher energy level within the atom. However, if the added energy is not high enough, the electrons will not change levels.

2 LIGHT RADIATION SOURCES

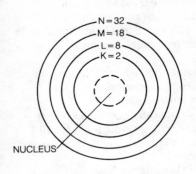
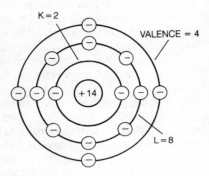

a. Maximum Number of Electrons in Each Shell   b. Shell Structure of Silicon

***Figure 2-6.** Shell Structure of Atoms*

When an electron does change energy levels, the energy level which lost the electron needs an electron to complete its shell and develops forces to try to attract an electron. It may capture the original electron it lost or it may capture an electron from added external energy. In either case, when it does capture another electron, energy is released in the form of radiation. The wavelength of the radiation depends on the change in energy, which in turn depends on the atomic structure. In some materials, the radiation may be a relatively low-frequency broad-spectrum electrical noise. In other materials, it may be visible light. If visible, the color of the light depends on the wavelength of the radiation. Some of the radiation may be absorbed by the material itself and simply cause the temperature of the material to increase. In any case, simply stated, when external energy is added, electrons are raised to a higher level; when electrons fall from the higher level to the lower level, light radiation is released.

Luminescence

The word *luminescence* was introduced by E. Wiedemann around 1889 to describe light emission which is not caused solely by the temperature of the material. Luminescence may be classified by the kind of excitation energy that produces the luminescence. Some of these classifications are chemiluminescence, bioluminescence, and cathodoluminescence.

Chemiluminescence occurs when excitation is supplied by a chemical reaction. The chemical reaction directly provides the necessary energy to produce light. An example is phosphorus glowing through oxidation in air.

Bioluminescence is actually a subdivision of chemiluminescence and occurs in living organisms. Examples are fireflies and glowworms.

LIGHT RADIATION SOURCES 2

Cathodoluminescence occurs when excitation is supplied by an accelerated electron colliding with atoms. This causes an electron in the atomic structure to move from one orbit to another which produces light. An example is the cathode ray tube used in television receivers, video terminals, and oscilloscopes to display information.

Luminescent sources may also be classified according to the way the source material generates light. Some of the common classifications are fluorescence, phosphorescence, and injection luminescence.

In fluorescent sources, applied energy raises an electron to a higher energy level. The electron remains in the higher energy state for about 10 nanoseconds (10^{-8} second) and then returns to its original energy level. As it returns to its original enery level, light is emitted.

In phosphorescence the electrons are in the intermediate state above their basic energy levels. The electrons are relatively stable but gradually fall to their basic energy levels. These materials continue to emit light even after the excitation source is removed. Many analog watches use phosphorescent material on the dials and hands so that they are visible in the dark.

Injection luminescence occurs in semiconductor diodes under certain operating conditions. When the diodes are operated in the forward direction, the junction region is enriched with electrons and holes. As these two kinds of charge carriers recombine with one another, an electron is transferred from the conduction band to the valence band at every recombination. Energy equal to the difference between the energy level of the conduction band and the valence band is given up. Some of this energy is converted to light.

Incandescent Light Sources

The word incandescence means light emission due to the temperature of the source. In a broad sense, all objects are incandescent sources because all objects in the presence of radiant energy both receive and emit radiant energy. (Remember that "light" includes infrared and ultraviolet rays.)

The tungsten filament lamp like those used in our houses *(Figure 2-7)* is the most common example of an incandescent source. During manufacture the glass envelope is flushed with an inert gas, evacuated, and hermetically sealed to prevent oxidation of the tungsten filament at high temperatures. The size and length of the tungsten filament is selected to obtain different power ratings and voltage ratings. Heating of the tungsten filament is caused by collisions of the electrons from the electrical supply with the molecular structure of the tungsten.

2 LIGHT RADIATION SOURCES

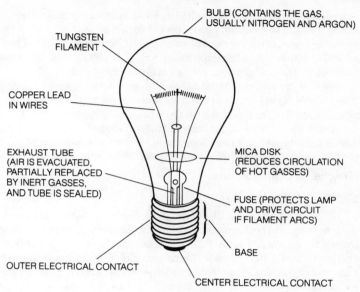

Figure 2-7. Tungsten Filament Lamp

The total radiant power varies with the filament temperature as shown in this equation:

$$\text{Total Radiant Power} = \text{Constant} \times (\text{Temperature})^4$$

Therefore, if the temperature increases by 10%, the total radiant power increases by 46%. If the temperature doubles, the total radiant power increases by a factor of 16. The temperature commonly used for the filament ranges from 2000° to 3500° Kelvin. (Degrees Kelvin is a temperature reference to absolute zero temperature. Temperatures given in °C can be converted to °K by adding 274 to the °C value.)

The light output from a tungsten filament bulb varies with voltage as shown in this equation:

$$\text{Light Output} = \text{Constant} \times (\text{Voltage})^{3.5}$$

Therefore, if the voltage increases by 10%, the light output increases by 39%. If the voltage doubles, the light output increases by a factor of 11.3.

Higher filament temperatures caused by higher voltage reduces the life expectancy of the lamp. In fact, the life varies with the applied voltage approximately as shown in this equation:

$$\text{Life (hours)} = \frac{\text{Constant}}{\text{Voltage}^{12}}$$

LIGHT RADIATION SOURCES 2

Therefore, if a lamp rated at 120 volts has a life expectancy of 1000 hours, then operating it at 130 volts would produce a life expectancy of only 382 hours. If the same lamp were operated at 110 volts, the life expectancy would be 2840 hours; and at 100 volts the life expectancy would be 8900 hours. Of course, the light output will decrease as the voltage is lowered.

Because of its versatility the incandescent lamp has many applications, but it has characteristics that prevent its use in some applications. For example, the switching time from off to on takes about one second. This is much too slow to use for rapid transfer of information. Therefore, information displays which require fast switching times require another type of light source such as the gas discharge lamp or a semiconductor light source.

Gas Discharge Light Sources

Fluorescent Lamp

The most common gas discharge light source for household use is the fluorescent lamp. It is constructed in a tubular form which may be straight or circular. As shown in *Figure 2-8a*, the glass enevlope is coated with a fluorescent material on the inside. During manufacture, air is withdrawn from the tube and replaced with an inert gas (often argon) and mercury vapor, and the evacuation hole is sealed. A heater and cathode assembly is located at each end of the tube. The heater filament is usually tungsten and the cathode is coated with a material that gives off electrons when hot. A base fits over each end and provides contacts to connect the heater and cathode to the external circuit.

The fluorescent lamp is typically connected in a circuit like the one shown in *Figure 2-8b* . Depending on the type of lamp, the start switch may be an external switch that must be held until the cathode is hot or it may be automatically controlled. When the start switch is closed, a large value current flows through the ballast and filaments. This stores an inductive charge in the ballast and heats the filaments, which indirectly heat the cathode. The cathode emits electrons which ionize the hot gas around the cathode at both ends. (An ionized gas is one that has been charged either positively or negatively.) Then the start switch is released and the high voltage produced by the inductive charge in the ballast causes an electrical arc (like a spark) to flash from one end of the tube to the other. This ionizes the gas throughout the tube and establishes a low resistance path for current flow through the tube. The ballast now serves as a current limiter to prevent excessive current flow. The electrons flowing through the tube collide with gas molecules and maintain the gas in an ionized state. Individual atoms, however, are constantly increasing and decreasing in energy level and ultraviolet light is emitted by the gas. Since ultraviolet light is invisible, it is of no value for illumination. That's why the inside of the tube is coated. The phospors in the coating are excited by the ultraviolet light and they emit visible light with a wide frequency spectrum that is almost white light.

2. Light Radiation Sources

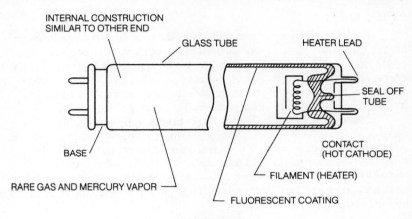

a. Lamp Construction

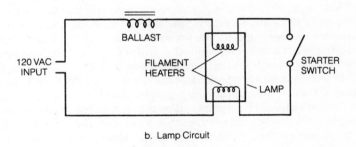

b. Lamp Circuit

Figure 2-8. Fluorescent Lamp

Neon Lamp

Another common gas discharge lamp is the neon lamp which is used for commercial outdoor signs, overvoltage protection, voltage regulation, sources for photoresistive devices, indicators, and illuminators. While the basic operating principles are the same as for the fluorescent lamp, the neon lamp doesn't require a starter circuit or phosphor coating. A high voltage is applied directly to the electrodes to strike the initial arc to ionize the neon gas. A current limiting device is used. The light emitted by the neon gas is red-to-orange so the glass envelope is clear to allow the light to pass. The color of the light can be changed to blue if a small percentage of mercury is added to the neon.

LIGHT RADIATION SOURCES 2

Xenon Lamp

The xenon lamp operates similarly to the neon lamp but produces an intense blue-white light. It is often used in strobe lights that produce a short, bright flash such as for photography and timing lights. Gas discharge lamps are capable of switching in about 0.5 second. Another gas discharge light source that produces a very bright light is the mercury vapor lamp that is frequently used for street lights and security lighting.

The wavelength or color of the light produced by a gas discharge lamp depends on the type of gas and the gas pressure. Low pressure gas discharge lamps tend to produce light with several discrete wavelengths while those using high pressure gas tend to produce light with a broader spectrum of wavelengths similar to incandescent lamps. Besides those already mentioned; mercury produces green light; carbon dioxide produces white light; and helium produces purplish-white light.

Semiconductor Light Sources

As was mentioned earlier, a semiconductor diode can be made to emit light. This is a result of the hole-electron recombinations that take place near the junction of a forward biased PN junction. The phenomenon is called electroluminescence or injection luminescence. If the light given off is visible, the device is called a visible light emitting diode (VLED). If the light is not visible, the device is called a light emitting diode (LED).

When electrons are injected into the N region of a PN diode as shown in *Figure 2-9*, and are swept through the region near the junction, they recombine with a hole. This recombination is a phenomenon similar to an electron in a high energy state returning to a lower energy state of the atom. As was discussed earlier, any time an electron interacts with an atom in this manner, light is generated at a frequency determined by the difference in the energy levels. In order for this recombination to result in luminescence of the material, there must be a net change in the energy levels and the photon generated must not be recaptured in the material.

The quantum efficiency of conventional light emitting diodes (photons out divided by electrons in) is a small percentage but is steadily increasing with advancements in technology. The light output power efficiency (light output power divided by electrial input power) is also low—less than 1%. It should also be noted that as temperature increases, the efficiency decreases because of an increase in nonradiative recombinations. Most light emitting diodes also decrease in efficiency as they age. In spite of these disadvantages, however, these devices are very useful as indicator lamps, pulsed sources, and illumination of small areas because of their long life, low cost, and fast response time. Switching speeds in the nanosecond range are available.

2 LIGHT RADIATION SOURCES

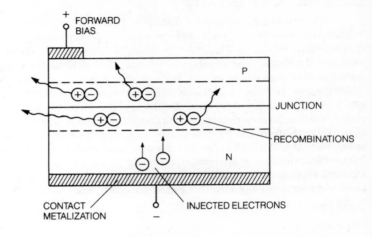

Figure 2-9. Cross Section Diagram of Light-Emitting Diode

Table 2-3 shows semiconductor materials with their typical wavelength and radiation range. The most commonly used semiconductor sources are gallium arsenide, gallium arsenide phosphide, and gallium phosphide.

***Table 2-3.** Semiconductor Light Source Materials and the Wavelength of Their Light*

Material	Wavelength (Å)	Radiation Range
Indium Antimonide	69000	Infrared
Indium Arsenide	34500	Infrared
Gallium Antimonide	17700	Infrared
Indium Phosphide	9850	Infrared
Gallium Arsenide	8980	Infrared
Gallium Arsenide Phosphide	6500	Red
Gallium Phosphide	5650	Green
Gallium Nitride	4000	Violet

LIGHT RADIATION SOURCES

2

Control of Reflected Light—LCD

It is important to recognize that in the field of optoelectronics it is possible to control not only the source and the detector, but also the transmission medium. A VLED device displays information to the eye by control of the source. The liquid crystal display (LCD), though serving the same function, does not emit light. It simply controls the medium through which the light travels. This is achieved by use of two polarization plates which sandwich a layer of liquid crystals. A voltage applied between the polarization plates changes the direction of polarization of the liquid crystals so that light is either reflected from the crystals or allowed to pass through. By controlling selected areas of the liquid crystals, characters can be formed to display information. One strong advantage is that very little power is consumed. Disadvantages are slower switching speed than the VLED, a small temperature range over which operation is reliable, and lack of visibility in the dark.

Lasers

Until 1954, electromagnetic radiation was generated and controlled using free electrons that moved from one atom to another. In 1954, a new concept was developed which utilized the control of energy states within atoms to produce electromagnetic radiation. One of the first such devices produced frequencies near 24,000 megahertz which is in the microwave frequency range. The device was called a maser (Microwave Amplification by Stimulated Emission of Radiation). In 1960, a maser capable of producing frequencies in the optical range was developed and demonstrated. It was called a laser (Light Amplification by Stimulated Emission of Radiation). The term optical maser also refers to a laser.

The principal characteristic of laser light is that the light rays are in phase (coherent), traveling in the same direction, and essentially of the same wavelength or color (monochromatic). As a result, the laser beam does not diverge a significant amount as it moves through air and maintains a high energy density. (By contrast, light from an incandescent lamp, even after being focused into a beam such as in a spot light, diverges rapidly as it travels through air.)

To explain the operation of the laser, it is more appropriate to consider light to be composed of discrete packets of energy called photons. These photons are uncharged particles which have an energy that depends only on their frequency (or wavelength). The amount of energy in the photon is given by this equation:

$$E = (4.137 \times 10^{-15})f \qquad (2\text{--}1)$$

where E is the energy in electron volts and f is the frequency in cycles per second or Hertz. The equation simply states that the energy contained by a photon of light is directly related to frequency; that is, energy increases as frequency increases.

2 LIGHT RADIATION SOURCES

In an atom, electrons orbit the nucleus at specific energy levels. An electron can move from a lower to a higher energy state only if additional energy is provided; that is, if an electron is at an energy level E_1, it can move to a higher energy level E_2 only if it receives an additional energy of E_2-E_1 as shown in *Figure 2-10a*. A photon of light with a frequency of

$$f = \frac{E2 - E1}{4.137 \times 10^{-15}}$$

could cause the electron to move to the higher energy level E_2. When the electron falls back to energy level E_1, as shown in *Figure 2-10b*, it emits a photon of light with the same frequency as the original. This is called *spontaneous emission*.

If an electron is already in the higher state, a photon of the proper frequency stimulates the electron at the higher energy level E_2 to fall to level E_1 and emit a photon in phase with the photon that stimulated it so that one photon in produces two photons out as shown in *Figure 2-10c*. This is called *stimulated emission*.

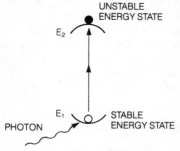
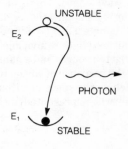

a. Photon raises electron from stable to unstable energy level (absorption)

b. When electron falls back to stable state, photon is emitted (spontaneous emission)

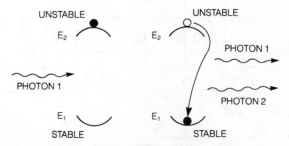

c. Photon 1 stimulates electron that is in unstable level causing stimulated emission

***Figure 2-10.** Photon Emission*

LIGHT RADIATION SOURCES 2

The higher energy level is called the *excited state*. If a large percentage of the atoms are in the excited state, it is said to have a *population inversion*. If a population inversion exists, the probability of stimulated emission of photons improves. The process of creating the population inversion is sometimes called *pumping;* that is, energy from some source must be "pumped" into the laser to provide the additional energy to create the population inversion. Then a photon of the proper frequency can create spontaneous and stimulated emission of two photons, these two photons stimulate four photons; and as the process continues, the intensity of the light increases while the phase and wavelength stay the same.

The actual construction of the laser depends on the type of material to be used. Many materials can be used to act as lasers or masers but the most common types are helium-neon, ruby, semiconductor, and carbon dioxide.

WHAT ARE THE IMPORTANT CHARACTERISTICS OF LIGHT SOURCES?
Spectral Distribution

The spectral distribution of a light source is the relative light output versus wavelength. This is an important consideration in matching the detector to its application or detector. For example, if the purpose of the lamp is to illuminate a display for people, an infrared source would be a poor choice since infrared light is invisible. *Figure 2-11* shows the relative spectral output of a tungsten source and a gallium arsenide PN LED (TIL31) source. It also shows the response of silicon phototransistors and the human eye. More detail on matching the spectral output of a source to the spectral response of a detector will be presented in Chapter 3.

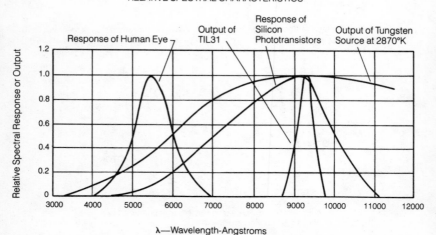

Figure 2-11. *Spectral Response Curves*

2 LIGHT RADIATION SOURCES

Viewing Angle

Most light sources do not emit uniformly in all directions. The most revealing directional information about a light source is its relative output versus the viewing angle displacement from the optical axis. *Figure 2-12* shows typical curves for the TIL31 and TIL33 LEDs. Reflectors, diffusers, or lenses may be used to control the viewing angle. For example, the TIL33, which has a flat lens, has a wider viewing angle than the TIL31 which has a domed lens.

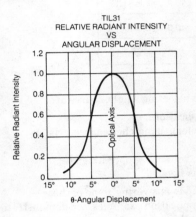
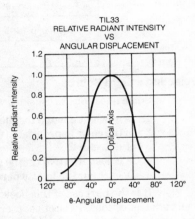

Figure 2-12. Viewing Angle Curve for TIL31 and TIL33

Efficiency

Efficiency is the ratio of power or luminous intensity out of a device to the power input to the device. In the radiometric system, efficiency is the ratio of radiant power out to electrical power in. In the photometric system, efficiency is the ratio of the visible power out in lumens to the electrical power in. Some typical efficiencies in the photometric system are shown in *Table 2-4*.

LIGHT RADIATION SOURCES — 2

Table 2-4. *Light Source Efficiencies*

Light Source	Efficiencies
Edison carbon lamp	1.8 lumens/watt
Tungsten lamp 1907	7.8 lumens/watt
Tungsten lamp 1950	13.9 lumens/watt
Xenon lamp	40.0 lumens/watt
Fluorescent lamp	70.0 lumens/watt

Selecting a Light Source

Some questions to consider when selecting a light source are:
1. Is it bright enough for the application?
2. Is the color appropriate (wavelength matches characteristics of detector)?
3. Is it cost effective (initial cost, reliability, operating cost)?
4. Does it put the light where it is needed?

It is the purpose of this book to aid in answering some of these application questions. In later chapters, application examples will demonstrate the use of manufacturer's data sheets and other information to select a light source and other components of an optoelectronic system.

WHAT HAVE WE LEARNED?

1) The major source of natural radiant light is the sun.
2) The majority of our energy sources originally came from the sun. The solar energy was converted to other forms of energy and stored.
3) Specialized light sources are developed to meet requirements of safety, portability, efficiency, reliability, and ease of operation.
4) Light can be produced by incandescence, luminescence, gas discharge, or electroluminescence.
5) Light is emitted from a source when electrons in its atomic structure are stimulated to vibrate.
6) Many energy sources may be used to make an electron vibrate in an atomic structure and emit light.
7) Light sources may be characterized by their frequency spectrum.
8) Applications of light sources require a knowledge of efficiency, intensity, viewing angle, and spectral distribution.

2 Light Radiation Sources

Quiz for Chapter 2

1. The principal source of natural light is
   a. the moon.
   b. the stars.
   c. the sun.
   d. oil.

2. Light sources continue to evolve toward
   a. portability.
   b. high efficiency.
   c. reliability.
   d. all of the above.

3. An incandescent light source using a tungsten filament produces visible light at a filament temperature of about
   a. 0°C.
   b. 100°C.
   c. 3000°K.
   d. 100°K.

4. Gas discharge light sources are
   a. used for voltage regulation.
   b. more efficient than incandescent lamps.
   c. reliable.
   d. all of the above.

5. Which of the following is a gas discharge source?
   a. neon
   b. fluorescent
   c. xenon
   d. all of the above

6. Reflective sources are important because they
   a. represent a controlled medium.
   b. emit light.
   c. store light.
   d. are visible in the dark.

7. Laser is an acronym for
   a. microwave.
   b. wavelength.
   c. light amplication by stimulated emission of radar.
   d. light amplification by stimulated emission of radiation.

8. Laser light is
   a. emitted uniformly in all directions.
   b. emitted at all frequencies.
   c. monochromatic.
   d. incoherent.

9. When a photon enters a laser material, it may
   a. be absorbed (by raising an electron to a higher level).
   b. stimulate the emission of an identical photon.
   c. not interact with the material
   d. all of the above.

10. A light source chosen for illumination for human use should have
    a. wavelengths greater than 9300Å.
    b. wavelengths less than 300Å.
    c. wavelengths between 4000Å and 7000Å.
    d. only the wavelength 5400Å.

11. The life expectancy of a tungsten filament lamp
    a. increases as voltage increases.
    b. decreases as voltage increases.
    c. increases as temperature increases.
    d. none of the above.

12. The radiant power of a tungsten filament lamp
    a. doubles if the temperature doubles.
    b. decreases as temperature increases.
    c. increases by a factor of 16 if the temperature doubles.
    d. none of the above.

13. In the shell structure of an atom, the outer shell is called
    a. the K shell.
    b. the nucleus.
    c. the valence shell.
    d. the N shell.

Understanding Optronics

LIGHT RADIATION SOURCES 2

14. The maximum number of electrons in the outer shell of any atom is
   a. 2
   b. 4
   c. 8
   d. 16

15. Semiconductors have _____ valence electrons
   a. 2
   b. 4
   c. 8
   d. 16

16. The word used to describe light produced by fireflies is
   a. chemiluminescence.
   b. bioluminescence.
   c. cathodoluminescence.
   d. incandescence.

17. In a semiconductor PN junction, light is produced when an electron recombines with a hole at the
   a. contact with the P material.
   b. contact with the N material.
   c. PN junction.
   d. all of the above.

18. The most efficient display for a watch is
   a. VLED.
   b. LCD.
   c. neon.
   d. incandescent.

19. A special starting circuit is required for
   a. LED lamps.
   b. LCD displays.
   c. incandescent lamps.
   d. fluorescent lamps.

20. If it is necessary to switch a light on and off very rapidly, the best choice for the light source is
   a. LCD.
   b. LED.
   c. incandescent.
   d. fluorescent.

1. c, 2. d, 3. c, 4. d, 5. d, 6. a, 7. d, 8. c, 9. d, 10. c, 11. b, 12. c, 13. c, 14. c, 15. b, 16. b, 17. c, 18. b, 19. d, 20. b

3 LIGHT DETECTORS

Light Detectors

WHAT IS A LIGHT DETECTOR?

A light detector is anything which responds to light energy; but in order to be of value, the detector must convert the light energy to another form of energy which is useful. For example, the eye converts light to electrical signals which are processed and interpreted by the brain to produce the sense of sight. A solar collector, such as a solar water heater or greenhouse, converts light energy into heat. Solar cells convert light into electricity which may be used immediately or stored in batteries for future use. In plants which contain chlorophyll and related pigments, photosynthesis uses light to convert carbon dioxide and water into the organic matter of which the plant is made.

In these examples, we see that the conversion process directly produces useful forms of energy from light. Other processes use light energy to control energy sources rather than produce energy. For example, light energy controls the resistance of a photoresistor which in turn controls electrical energy. Other examples are photodiodes, phototransistors, and photo SCR's which are controlled by light energy and in turn may be used to control electrical power.

Natural Light Detectors

In the broad sense, everything is a light detector because any object will convert sunlight to heat. However, most of these objects are not particularly useful as detectors because they give very little information about the characteristics of the light or the medium through which the light came. That is, we might like to know the direction from which the light came, the intensity of the light, and the wavelengths or color of the light. These are hard to determine from the temperature of an object.

In contrast, the most exotic detector is a natural detector—the eye. The eye provides enough information about light to allow us to determine the direction of the light source, to establish relative light intensity, and to distinguish between wavelengths or color of light. The characteristics of the eye as a detector vary among humans, animals, and insects depending on their specialized needs. Bees have segmented eyes which can be used for navigation by establishing a pattern that depends on the sun's position. Eagles have extremely good long distance eyesight which allows them to detect motion from great distances. There are many other examples of specialized adaptation

LIGHT DETECTORS 3

in the animal kingdom which are essential to the animals' survival. The same thing can be said for humans but we usually want much more information than that required only for survival. In the effort to extend our ability to obtain information, we have developed specialized light detectors to replace the eye in routine activity, to obtain and store information in our absence, and to provide information that our eyes cannot see.

Specialized Light Detectors

Although light has been used since time began, the control of light energy was limited to the storage of energy in the form of heat in various forms of solar collectors (rocks, water, etc) and the use of simple optics (mirrors, lenses, telescopes, and microscopes) until relatively recent time. With the discovery and understanding of electrical energy, more sophisticated light detectors came into use.

The first large scale commercial development which motivated the need to know more precise details about light intensity and wavelength was the development of photosensitive films and the resultant development of the camera. Thus photographic film was one of the first of the modern light detectors. Photography also motivated other developments. The film required different exposure times depending upon the intensity of the light and characteristics of the film. Early films required trial and error approaches which motivated the development of reliable light meters to reduce the chances of over- or under-exposing the relatively expensive film.

The development of the camera is also a good example of the impact of optoelectronics. A camera requires a light source, a transmission medium, and a detector. The light source which illuminates the subject must have enough intensity to reflect a sufficient amount of light to the detector (film). The light source must contain a broad enough spectrum of wavelengths to allow the reflective properties (such as color) of the object to be determined and accurately recorded on the film. The transmission medium (lens) must allow adequate intensity of the reflected light to reach the film and must properly focus the image on the film. The detector (film) must respond differently to the various wavelengths and must be capable of providing the contrast and resolution necessary to record the information contained in the reflected light. The requirements of all three parts of this photographic system interact to determine the final system. For example, the characteristics of the film (detector) dictate the exposure time as a function of the reflected light intensity which is transmitted through the lens. The position of the object relative to the film dictates focal length, the adjustment of the lens for correct focus, and the light source intensity.

3 LIGHT DETECTORS

Early photographers (*Figure 3-1*) looked through the camera lens at the subject and estimated the required exposure time based on their knowledge of the film's characteristics. Lenses were adjusted to establish the proper focus, the film was placed into the camera, and the shutter was opened for the length of time estimated by the photographer. By contrast, many modern cameras (*Figure 3-2*) are completely automatic with automatic focusing, exposure control, aperture control, film advance, flash control, as well as more status indicators than the average person can use. All of these modern features have been accomplished with specialized light detectors that produce or control electrical energy.

Figure 3-1. *Early Photographer*

Figure 3-2. *Canon AF35M Sure Shot Camera*
(Courtesy of Canon, Inc.)

LIGHT DETECTORS 3

HOW DO LIGHT DETECTORS WORK?

Specialized light detectors can be grouped into two broad categories; thermal detectors and quantum detectors. Thermal detectors convert light energy to heat which is in turn used to produce another form of energy, usually electrical.

Quantum detectors are commonly divided into three subgroups; photoresistive, photovoltaic, and photoemissive. Photoresistive detectors use light energy (photons) to control the resistance of the detector. Photovoltaic detectors use light energy to produce an electrical voltage. Photoemissive detectors use light energy to free electrons from the detectors surface to produce a current.

How Do Thermal Detectors Work?

When current flows through a material that has electrical resistance, electrical power is converted to heat. This is an irreversible process; that is, heating the material will not produce electrical power.

There is a reversible process, called the Thomson effect, which can be used to convert heat to electrical energy. The Thomson emf is produced when one end of a metal bar is heated. The heat energy increases the activity of the free electrons in the hot end of the bar and, in a sense, increases the pressure to force electrons toward the colder end of the bar. This creates an imbalance of electron distribution in the bar and produces a potential difference or voltage in the millivolt range between the ends of the bar.

Another effect, called the Peltier emf, exists when two different metals are joined. A diffusion of electrons from one metal to the other builds up an electromotive force (emf) in the order of millivolts at the junction. The Thomson and Peltier effects can be used together to produce the Seebeck emf which is used in thermocouples. Two different metals are connected to form a closed circuit as shown in *Figure 3-3*. If the junctions are at different temperatures, there will be a Thomson emf in both metals. Since the metals are different there will be a Peltier emf at each junction but since the junctions are at different temperatures, one of the Peltier emfs is larger than the other. The result is a net emf resulting from adding the Thomson and Peltier emfs around the closed circuit. This net emf produces a current in the circuit which is proportional to the temperature difference between the two junctions. This device is called a thermocouple and is commonly used in natural gas furnaces and water heaters as a safety device. When the pilot light is on as shown in *Figure 3-4a*, it heats the thermocouple which produces a current sufficient to hold a solenoid valve in the main gas line open. If the pilot light goes out as shown in *Figure 3-4b*, the thermocouple current decreases and the solenoid valve shuts off the main gas supply to the heater.

3 Light Detectors

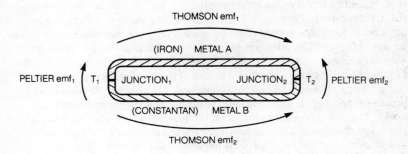

Figure 3-3. Construction of Thermocouple

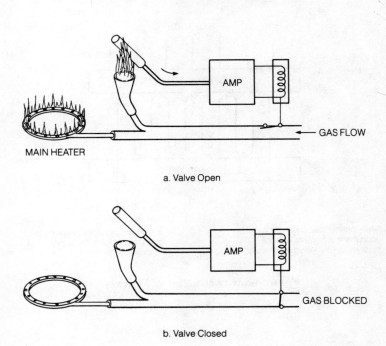

Figure 3-4. Thermocouple Controlling Gas Flame

LIGHT DETECTORS 3

Several thermocouples connected in series, as shown in *Figure 3-5*, to increase the emf is called a thermopile. The thermopile can be used as a radiation receiver, a sensitive detector of radiant energy. A radiation receiver has a set of junctions in good thermal contact with the receiver but electrically insulated from it. The other set of junctions are attached to a support which does not receive the radiation and, therefore, is at a lower temperature. The incident radiation raises the temperature of the receiver and produces a voltage proportional to the energy absorbed. The thermopile responds to a broad spectrum of optical wavelengths both in the visible and nonvisible range. The speed of response, that is, the rate with which the thermopile can follow changes in light intensities, depends largely on the thermal characteristics of the receiver. It is important to remember that the thermopile voltage is proportional to a temperature difference and is; therefore, proportional to the total radiation energy received.

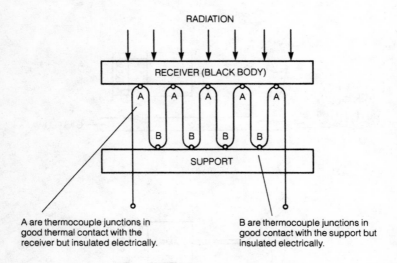

A are thermocouple junctions in good thermal contact with the receiver but insulated electrically.

B are thermocouple junctions in good contact with the support but insulated electrically.

Figure 3-5. *Construction of Thermopile*

3 Light Detectors

Review of Semiconductor Materials

Before proceeding with the discussion of quantum light detectors, let's briefly review how semiconductor materials are made.

A diagram of the shell structure of a silicon atom is shown in *Figure 3-6a*. In the simplified diagram of *Figure 3-6b*, the four lines represent the four valence electrons of silicon which are used in forming the covalent bond with other atoms.

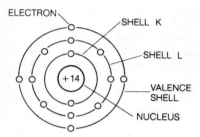

a. Shell structure of silicon showing all shells with electrons and corresponding protons as a positive number in the nucleus.

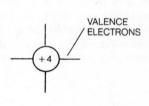

b. Simplified diagram for silicon showing only outer shell (valence) electrons as radial lines and corresponding protons as a positive number.

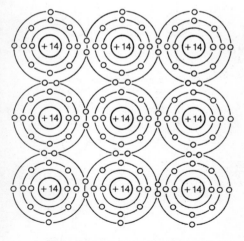

c. Silicon atoms sharing valence electrons with all shells shown.

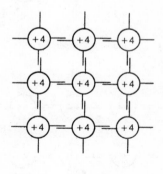

d. Silicon atoms sharing valence electrons using simplified method for diagram.

***Figure 3-6.** Silicon Shell Structure*

LIGHT DETECTORS 3

Groups of atoms with four valence electrons join together with a covalent bond. In the covalent band, valence electrons are shared by adjacent atoms to fill the valence band of each atom with eight electrons to form a stable structure. *Figure 3-6c* and *3-6d* illustrates a group of silicon atoms sharing valence electrons so that each atom has access to eight electrons to form covalent bonds. *Figure 3-7a* again shows the silicon covalent bonds for comparison to *Figure 3-7b* which illustrates a number of atoms with three valence electrons (gallium or indium) and atoms with five valence electrons (phosphorus, antimony, or arsenic). These materials are called Group III and Group V semiconductors. In this combination, the "extra" electron from each group five atom fills in the vacancy in the covalent bond (eight electrons required) created by the group three atom. The result is a material that behaves much like silicon or germanium (another material with four valence electrons) as a semiconductor material.

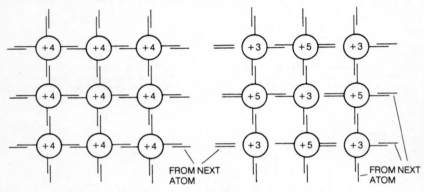

a. Silicon Covalent Bonds

b. Covalent bonds formed by equal proportions of group 3 and group 5 atoms

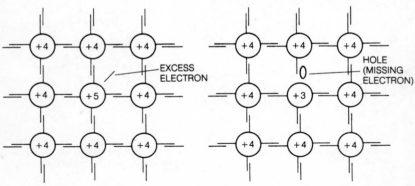

c. N-Type Semiconductor Material

d. P-Type Semiconductor Material

***Figure 3-7.** Structure of N-Type and P-Type Materials*

3 LIGHT DETECTORS

Several other combinations of materials which behave as semiconductor materials are possible. Some of these combinations are gallium phosphide, indium arsenide, and gallium arsenide phosphide. These combinations are important because these materials may be used as visible light emitters.

If an atom from group five (called a donor atom) is added to the structures of either *Figure 3-7a* or *3-7b*, the extra electron is not used in the covalent bond as shown in *Figure 3-7c*. This extra electron is called a "free" electron in the sense that there are no strong valence forces holding it in place. This electron is available as a negative charge carrier and moves easily through the material. The material is referred to as an "N-Type" (negative excess charge) semiconductor material.

If an atom from group three (called an acceptor atom) is added to the structures of *Figure 3-7a* or *3-7b*, a vacancy is created in the valence structure as shown in *Figure 3-7d*. This vacancy is called a "hole". Since the hole is due to the absence of an electron (negative charge) the hole can be considered to be a positive charge carrier. The material is refered to as a "P-Type" (positive excess charge) semiconductor material.

How Do Photoresistive Detectors Work?

Photoresistive detectors are based on the fact that external energy added to a material can remove an electron from its parent atom. The free electron becomes available as a current carrier and reduces the effective resistance of the material. The energy required to free the electron is called the energy gap and depends on the type of material as shown in *Table 3-1*.

Table 3-1. Energy Gap for Some Materials

Name	Chemical Symbol	ev at 300°K[1] Optical Energy Gap
Cadmium Sulfide	CdS	2.4
Gallium Phosphide	GaP	2.2
Cadmium Selenide	CdSe	1.7
Gallium Arsenide	GaAs	1.4
Silicon	Si	1.1
Germanium	Ge	0.7
Indium Arsenide	InAs	0.43
Lead Sulfide	PbS	0.37
Lead Telluride	PbTe	0.29
Lead Selenide	PbSe	0.26
Indium Antimonide	InSb	0.23

[1]Degrees Kelvin is a temperature reference to absolute zero temperature.
Common temperatures in °C can be converted to °K by adding 274 to the °C value.

LIGHT DETECTORS

3

The relationship between energy and wavelength of light is given in this equation.

$$E(ev) = \frac{12{,}400}{\lambda(\text{angstroms})} \quad (3\text{-}1)$$

where E = energy in electron volts (ev) and λ = wavelength in angstroms. Using equation *3-1* and *Table 3-1* it is seen that the energy necessary to generate a free hole-electron pair ranges from 0.2 to 2.4 electron volts which corresponds to radiation wavelengths between approximately 5,000 and 60,000 angstroms.

As shown in *Figure 3-8*, no junction is necessary to operate the photoresistive devices. A layer of photoresistive material is connected to two leads, and as the light intensity increases, the resistance between the leads decreases. Because this detector does not generate a voltage, an external voltage source must be used to make electrons flow through the detector and external circuit. Because the photoresistance decreases as light intensity increases, the current in the circuit increases as light intensity increases. The objective of using the detector is to convert light to electrical signals, so it is desirable to make the photoresistor of a material that does not reflect much light in order for a large percentage of the photons falling on the detector to produce hole-electron pairs. Typical construction is shown in *Figure 3-8b*.

From an electrical viewpoint, the most important characteristic of the photoresistive device is the dark to light resistance ratio, which is typically about 1000:1. Any other factors which may influence resistance must be considered in an application of a photoresistive detector. Temperature is usually the one that is of most concern because changes in temperature cause inherent resistance changes and can influence the production of hole-electron pairs. Therefore, both light intensity and temperature must be considered in applications, and in many cases the temperature may have to be controlled.

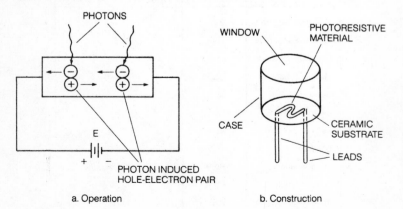

Figure 3-8. *Photo Resistive Device*

3-10 UNDERSTANDING OPTRONICS

3 LIGHT DETECTORS

How Do Photovoltaic Detectors Work?

Perhaps the most commonly used detectors in optoelectronic controls are the photovoltaic detectors—those that use the PN junction effects of semiconductor materials. These fall into several classes of devices which are commercially available. The photodiode is the simplest form and is the building block upon which photo transistors, photo FETs[2], photo SCRs[3], photo TRIACs[4], and many photo sensitive integrated circuits are built. A cross section of a reverse biased PN junction diode is shown in *Figure 3-9*. Notice that the depletion region of the PN junction acts as an insulation between the anode and cathode because free electrons and holes are not present in this region. When a photon with energy equal to the energy gap enters the depletion region, it is absorbed. As a result of this absorption, hole-electron pairs are formed. The hole-electron pair is separated and swept quickly out of the depletion region. The (−) portion of the pair will be moved to the cathode and the (+) portion to the anode. If the diode is biased as shown, the result will be a current flow through the reverse biased diode. If the hole-electron pair is produced outside the depletion region, then it will likely recombine and no photo current will be produced. Thus, to optimize the PN junction for use as a photodetector, the P region should be as thin as possible to reduce the probability of recombination while the depletion region should be as thick as possible to improve sensitivity. Another advantage of the thick depletion region is that it reduces junction capacitance. This permits the photodiode to be able to respond to fast changes in light levels.

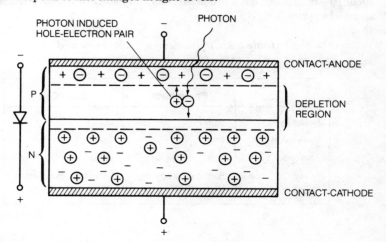

Figure 3-9. *Reverse Biased PN Junction*

[2] Photo Field-Effect Transistors
[3] Photo Silicon-Controlled Rectifiers
[4] Photo Silicon-Controlled Rectifiers Connected Head to Toe

LIGHT DETECTORS 3

If the diode shown in *Figure 3-9* is not biased, the continued exposure to light will produce a voltage across the diode. When the anode is connected to the cathode through a load resistance, a current proportional to the light intensity will flow through the external circuit. This means that the PN junction is capable of converting light energy directly to electrical energy. The PN junction device designed to perform this function is called a solar cell.

Since the purpose of a solar cell is to produce electrical power, it is desireable to reduce power losses, capture as much radiant energy as possible, and operate without external bias. These requirements result in a design which uses low resistivity material to reduce losses after the electrical power is generated and a large surface area to capture as much radiant energy as possible. Since the solar cell is not biased to remove the hole-electron pairs from the depletion region, it is necessary that the depletion region be thin to minimize recombinations in the depletion region. The large area and thin depletion region produce a high capacitance which causes a slow response to changes in light levels. The slow response makes the solar cell unsuitable for use as a photodetector in most applications. So we see that photodiodes and solar cells are very different although they are both based on the properties of a PN junction.

Some of the differences between photodiodes and solar cells are listed in *Table 3-2*. Solar cells have been used on virtually every satellite and space mission as a source of energy for the onboard electronic systems. Typical production solar cell conversion efficiency (ratio of power out to power in) is about 10% but research has produced laboratory solar cells with efficiency approaching 20%.

Table 3-2. *Comparison of Photodiode and Solar Cell Characteristics*

	Photodiode	**Solar Cell**
Resistivity	High	Low
Depletion region	Thick	Thin
Area of cell	Small	Large
Typical Operation	Biased (Photocurrent)	Unbiased (Photovoltaic)
Speed	Fast	Slow

3 LIGHT DETECTORS

How Do Photoemissive Detectors Work?

Photomultipliers are designed to multiply the very few electrons produced by the photoemissive effect of a low-level light signal in much the same manner that an audio amplifier amplifies a small signal from a microphone. One of these devices is the photomultiplier vacuum tube. A photomultiplier vacuum tube uses two effects, photoemission and secondary emission, that cause a material to emit or release electrons from the surface of the material.

As shown in *Figure 3-10*, photons striking the photocathode cause the release of electrons from the photocathode material through the process of photoemission. These electrons are focused into a beam by an electrostatic field and accelerated toward a curved plate called a dynode. As each high energy electron strikes the dynode, it adds enough energy to release two or more electrons from the dynode material. This process is called secondary emission. The secondary electrons are then accelerated toward another dynode where they produce even more secondary emission. This process may be repeated for several steps. The release of one electron from the photocathode caused by the photon may produce several thousand electrons as the secondary emission effect is multiplied through several stages. The overall effect can produce internal current gains from 10^5 to 10^8.

The quantum efficiency of a photocathode by itself ranges from 0.01 to 10 percent. This means that from 10 to 10,000 photons are required to produce one electron from the photocathode. By using the secondary emission effect, the photomultiplier vacuum tube improves the quantum efficiency so that from 10^3 to 10^7 electrons are produced for every photon.

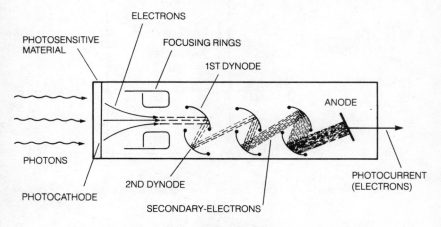

Figure 3-10. *Operation of Photomultiplier Vacuum Tube*

LIGHT DETECTORS   3

The effect of light on a photoemissive detector also may be multiplied using phototransistors or photoFETs. However, the semiconductor device which acts most like the photomultiplier vacuum tube is the avalanche photodiode. The avalanche photodiode operates in the avalanche breakdown region. Avalanche breakdown refers to a condition in which a PN junction is reverse biased to a point near the breakdown of the junction. That is the point at which the potential is so high that electrons can be pulled from the atomic structure. With a small amount of additional energy, electrons are dislodged from their orbits producing free electrons and resulting holes. The reverse bias condition widens the depletion region.

As illustrated in *Figure 3-11*, a photon with enough energy generates a hole-electron pair in the depletion region. Because of the action of the high electric field, the hole with its positive charge moves up toward the negative side of the battery, and the electron with its negative charge moves toward the positive side. The electron moving in the high electric field is accelerated and collides with other bound electrons. Because of the high velocity when the electron collides, additional hole-electron pairs are generated which are also accelerated. These may produce still other hole-electron pairs and an avalanche multiplication process results. Thus, one photon may produce up to 100 electrons in the avalanche photodiode.

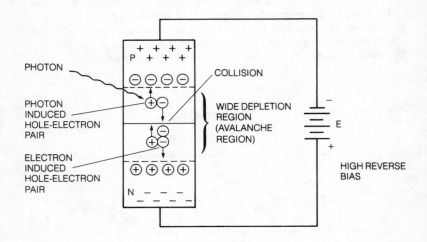

Figure 3-11. Operation of Avalanche Photodiode

3 LIGHT DETECTORS

WHAT ARE THE IMPORTANT CHARACTERISTICS OF LIGHT DETECTORS?

Several different types of detectors have been discussed. Each has different characteristics that provide advantages and disadvantages that make it functional or fit for a particular application. Consider the application of a detector used to switch on an outdoor light at dusk and off at dawn.

    a) What are the characteristics of the light source?

The source is sunlight which contains a broad spectrum of wavelengths with a peak intensity of about 80 mW/cm^2. The light should be turned on at dusk (after sunset) and turned off as soon as it is light enough to see (before the sun is visible). The light will come from a different direction at any given time.

    b) What characteristics must the detector have?

Since the light to be detected is sunlight, the detector must respond to wavelengths found in sunlight. This is the spectral response. It must be sensitive to light coming from a wide viewing angle. This is its viewing angle characteristic. In addition, the detector must produce an output large enough to be detectable at the low light levels available at dawn. This is its efficiency. All characteristics combine to make the system effective in turning the light on or off.

A second example which requires different characteristics is that of an intrusion alarm similar to the one shown in *Figure 3-12*. A lamp is positioned to shine across the door way and a sensor is positioned on the opposite side to sense the presence of light from the lamp.

    a) What are the requirements of the light source? What light source can meet the requirements?

For an intrusion alarm the light must not be visible. For this reason an infrared light source is ideal for this application. A gallium arsenide diode which emits infrared light at a wavelength of 9300 angstroms and has a light intensity such that 25 microwatts per square centimeter of light is delivered to the detector can be used as a light source.

LIGHT DETECTORS **3**

b) What characteristics must the detector have?

Let's take the same characteristics in the same order as the last example. First, what should the spectral response be? The detector must have a spectral response to produce an output when illuminated by the 9300 angstrom wavelength. Next consider the viewing angle. For this application a restricted viewing angle is needed. The viewing angle must be such that it responds to light coming from the direction of the source but does not respond to light from a different direction. Finally, the efficiency must be such that the detector produces a detectable change in the output with less than 25 microwatts per square centimeter illumination. When the light is interrupted the change in output from the detector is what signals the intrusion.

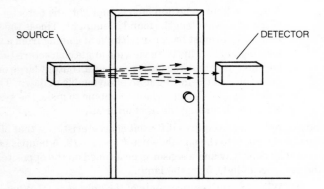

Figure 3-12. Intrusion Alarm

Spectral Response

Now let's examine the individual characteristics in detail. The spectral response of a detector is a measure of the response of the detector to the same quantity of light of different wavelengths. It is important to understand that a detector may sense light at one wavelength while it may not respond at all to that of another. For example, the eye can detect light in the visible range but cannot see a radio wave. Most data sheets for light detectors contain a graph similar to the one shown in *Figure 3-13* which displays the relative spectral characteristics of a phototransistor semiconductor detector, the TIL81, and the human eye. The human eye, as we discussed previously,

3 LIGHT DETECTORS

has a response to wavelengths from 4,000 to 7,000 angstroms. Notice that the peak response of the TIL81 is just above 9,000 angstroms. It is only half as responsive at 10,000 and 7,000 angstroms, and it hardly responds at all to wavelengths greater than 11,000 angstroms and less than 5,000 angstroms.

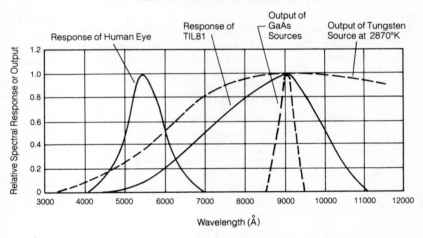

Figure 3-13. Relative Spectral Characteristics of TIL81 Detector

Notice that the spectral response is a relative spectral response. If absolute units are to be applied to the relative scale, a standard kind of source must be used. A common one is a tungsten filament bulb operating at 2870°K. Its light output is shown in *Figure 3-13* along with the response of the detectors. A common level of intensity of this tungsten filament bulb is 5mW/cm² ($E_e = 5mW/cm^2$). With this intensity the absolute output of the TIL81 will be greater than 5 milliamperes of light current (I_L) when 5 volts is used as the supply voltage. Since the advent of the gallium arsenide (GaAs) 9300Å infrared source, the most common reference source is now the infrared source. Its light output curve is also shown in *Figure 3-13*. Notice that the TIL81 has its maximum sensitivity at about 9300Å which matches the peak of the output of the TIL31 gallium arsenide source. Because of the match of spectral output of gallium arsenide sources with the spectral response of detectors such as the TIL81, these sources and detectors are often used together. The sources in that case are gallium arsenide light emitting diodes and the detectors are silicon photodiodes or phototransistors.

LIGHT DETECTORS **3**

Viewing Angle

The application examples considered above show that it is not only important to know the response of a detector to the spectrum of the light source, but also to know how that response changes with changes in the direction of the light source. *Figure 3-14* gives the normalized (relative) light current (I_L) output from the detector as the source is varied from the optical axis. The angular displacement is shown in *Figure 3-14*. Notice that the TIL81 has a peak response to light which comes directly into the device (source aligned directly on the optical axis) and is less than 20 percent as responsive to light coming into the device from a source located at angles greater than 20° off axis.

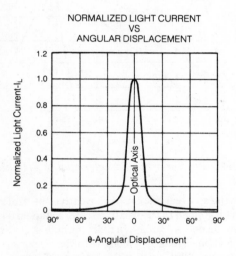

Figure 3-14. Viewing Angle Curve for TIL81

You can see then that the viewing angle is important because some applications require wide viewing angles when the light source is not in a fixed position with respect to the detector, or the detector may need to sense several sources in different positions.

A wide viewing angle is usually obtained at the expense of sensitivity. This effect is illustrated in *Figure 3-15*. The same chip is used with the same light source for both curves. With the lens, the viewing angle is reduced but more of the light is focused on the chip when the light source is aligned with the optical axis so that more light current is produced.

3 LIGHT DETECTORS

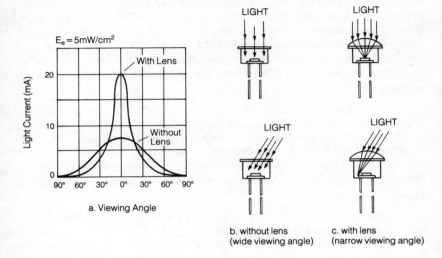

***Figure 3-15.** Comparison of Light Current for Device With and Without a Lens*

Efficiency

The word efficiency is normally defined as the ratio of the quantity of a desired effect to the effort required to obtain it. In most cases, the efficiency is the amount of output for a given input measured in the same units. For example, an electrical power transformer outputs 900 watts for every 1,000 watts in. Therefore, its efficiency is 90%; that is, 90% of the input that is supplied comes out as output. However, efficiency can also be expressed when input and output do not have the same units of measurement.

Suppose the desired effect is to move a person over a distance. Quite a few ways are available to do this, but a common way is to use an automobile. In an automobile the important efficiency is the fuel efficiency. Fuel efficiency is measured in miles per gallon—miles that the person is moved (the desired effect) divided by gallons (the effort required). The more miles the person is moved for a gallon of fuel the greater the fuel efficiency.

In the case of an electronic detector, the desired effect is light current and the effort required to produce it is light intensity. The more light current (the desired effect) for a given amount of light intensity (the effort required) the greater the efficiency.

LIGHT DETECTORS 3

The type of efficiency data given for a photodiode will vary from one manufacturer to another. Typical data that relates to efficiency includes a graph of diode current versus the energy density of the light (milliwatts per square centimeter). A typical value is 0.008 amperes per watt per square centimeter of active detector surface. The efficiency data is more useful for comparing devices than for determining absolute performance.

SUMMARY

The spectral response, viewing angle, and efficiency are the most important characteristics specified by the manufacturer. They are specified at standard test conditions. If an optoelectronic detector is to be used with conditions differing significantly from those specified by the manufacturer, it would be wise to establish the characteristics of the detector by testing it directly in the application. This is especially true of efficiency.

Electrical Properties

The electrical properties of detectors are important because we usually need to process the electrical output signal in some manner.

The thermocouple has a low internal resistance (tenths of an ohm) and a low output voltage (tens of millivolts). The power output depends on both the load resistance and the temperature. If it is necessary to deliver maximum power to a thermocouple load like a relay or solenoid valve, the relay or solenoid coil should have low resistance. In fact, for maximum power transfer, the coil resistance should be equal to the thermocouple's internal resistance.

The thermopile is normally used to provide an output voltage which is proportional to radiation. Its internal resistance is higher than a thermocouple (10 to 100 ohms) and its voltage is higher than a thermocouple (hundreds of millivolts). The load for the thermopile is usually chosen to have a resistance higher than the thermopile's internal resistance.

Photoresistive detectors are usually specified by the resistance versus light curves or the ratio of light resistance. These devices can be obtained with light resistance values from a few ohms to several thousand ohms and a dark to light resistance ratio of 1000:1. The power rating of photoresistive devices is usually in the milliwatt range. The power rating required can be calculated using the equation,

$$P_{rating} = \frac{V^2_{max}}{R_{min}} \qquad (3\text{-}2)$$

where V_{max} is the maximum voltage applied to the device and R_{min} is the minimum light resistance of the device.

3 LIGHT DETECTORS

The photodiode and solar cell both produce approximately 1.2 volts when operated in the photovoltaic mode. The solar cell has a low resistance (ohms) while the photodiode has a high resistance (hundreds of ohms). The solar cell should be used with low resistance loads for maximum power output while the photodiode can be used with higher resistance loads.

The electrical characteristics of the phototransistor, photo field-effect transistor, photo SCR and photo TRIACs are essentially the same as their corresponding non-photo detecting semiconductor devices.

Selecting a Photo Detector

Some questions to consider when selecting a light detector are:

1) Is it sensitive enough for the application?
2) Do the wavelength characteristic of the detector match the source?
3) Is it cost effective (initial cost, reliability, operating cost)?
4) Can it "see" the light source or sources?
5) Is it electrically compatible with the rest of the system?

In chapter seven, detailed applications of photodetectors will demonstrate the use of manufacturer's data sheets to help answer these questions.

WHAT HAVE WE LEARNED?

1) A light detector may be anything which responds to light energy.
2) The eye is the most sophisticated natural light detector.
3) Specialized light detectors made by people have been developed to assist us to detect and record information that we otherwise would not see.
4) Light detectors can be grouped into classes of thermal detectors or quantum detectors.
5) Thermal detectors convert heat to another form of energy (usually electrical) which is easily measured.
6) Quantum detectors either change resistance (photoresistive), produce a voltage (photovoltaic), or produce a current (photoemissive) in proportion to optical energy.
7) Photovoltaic detectors use the depletion region of a semiconductor PN junction to generate a voltage when stimulated by light.
8) Photomultipliers use an avalanche or secondary emission effect to amplify the effects of the light.
9) Important parameters of detectors include spectral response, viewing angle, and efficiency.

LIGHT DETECTORS 3

Quiz for Chapter 3

1. The purpose of a light detector may be to provide information about the
   a. light source.
   b. medium through which light travels.
   c. color of the light.
   d. all of the above.

2. A light detector may
   a. produce voltage.
   b. change resistance.
   c. multiply effect of light.
   d. any of the above.

3. A solar cell is a
   a. photovoltaic device.
   b. photoresistive device.
   c. photomultiplier.
   d. thermopile.

4. A thermopile responds to a
   a. narrow energy spectrum.
   b. broad energy spectrum.
   c. single color.
   d. low intensity only.

5. A photoresistor is made of
   a. a semiconductor PN junction.
   b. two different metals.
   c. bulk photoresistive material.
   d. photovoltaic materials.

6. Photodiodes make use of an electron-hole pair generated in the
   a. P-type material.
   b. N-type material.
   c. depletion region.
   d. cathode.

7. Photomultiplier devices increase the
   a. light.
   b. response to light.
   c. color.
   d. viewing angle.

8. All detectors have the same
   a. spectral response.
   b. viewing angle.
   c. sensitivity.
   d. none of the above.

9. A solar cell could be used as a photodiode if the application does not require
   a. high speed.
   b. large reverse bias.
   c. directional sensitivity.
   d. all of the above.

10. The photodetector with the broadest spectral response is the
    a. photodiode.
    b. photomultiplier
    c. photoresistor.
    d. thermopile.

1. d, 2. d, 3. a, 4. b, 5. c, 6. c, 7. b, 8. d, 9. d, 10. d

4 Optically Coupled Electronic Systems

Optically Coupled Electronic Systems

WHAT IS AN OPTICALLY COUPLED SYSTEM?

Light sources and light detectors are almost always used together. In fact, it can be argued that you cannot have a light source without a detector or vice versa. In this sense, all uses of light can be called optical systems since they involve more than one part.

An optically coupled system is used to obtain information or to transfer information from one point to another. We must understand the characteristics of the light source and the characteristics of the detector so that we can ensure that the detector and source are compatible. The detector must be able to "see" the source; that is, the detector must produce an output when illuminated by light from the source. The illuminating light must travel from the source to the detector, through some medium which we have called the transmission medium. It may be air, it may be outer space, it may be a vacuum, or it may be a material of one kind or another. For this reason, we can say that an optically coupled system consists of a light source, a light detector, and a transmission medium selected to work together as a system to perform the desired task.

Systems may have many variations. One source may be used that supplies many detectors. As we've discussed, the sun fits this bill. Living organisms all over the world depend on sunlight and the transmission of light energy. Other systems may have many light sources but only one detector. A scanned array used for facsimile system is an example of such a system. All are optical systems but a very common one is a system with one source and one detector separated by a transmission medium.

Natural Optically Coupled Systems

Still the most common system is one where a natural light source communicates information to the eye by the light reflected from our surroundings. Let's see what kind of information is transmitted. By observing the relationship between the movement of the sun, moon, and earth, we get information to measure time in hours, days, and years. Weather information is obtained from the look of the sky or the type and movement of the clouds. The color of flowers, the texture of material, the smiling face, the angry dog, the book page, the black of night, and the orchestra in the park are all pictures of information obtained via reflected light sensed and detected by the eye and interpreted by the brain.

Optically Coupled Electronic Systems 4

These reflected pictures can contain coded information. For instance, the American Indian used a fairly sophisticated natural optically coupled system when he sent smoke signals. The sun is the light source, the eye the detector, the light waves the carrier of the information, and the puffs of smoke the coded information. The transmission medium is the air.

Semaphore signals with flags, shuttered light for communicating between ships, and reflected sunlight from mirrors are other means of reflecting light or turning on and off light to provide coded information.

Now the characteristics of the source definitely effect the quality and character of the information received at the detector. Let's use the eye again as an example. Bright sunlight provides more information to the driver of an automobile than headlights at night because the sunlight is more intense, contains more wavelengths, and provides a greater viewing angle than headlights.

This example also demonstrates the influence that surroundings and the transmission medium have on the information received. Dark surfaces with white or yellow contrasting lines define the road much better for the night driver using headlights. Fog, rain, or dust cut down on the light from the source to objects of interest and even, as in the case of fog, reflect light back from the transmission medium to completely block any light being reflected from objects on the roadway.

Other examples probably come to mind. Have you ever walked into a photographic darkroom when the red light is turned on? The somewhat eerie sensation is due to the fact that less information is obtained by the eye than when natural light is being used as the source.

Specialized Optically Coupled Systems

Specialized optically coupled systems are all systems designed to do a specific job to solve a particular problem. Specialized optically coupled systems can generally be divided into two types; either *interruptible* or *non-interruptible*.

Interruptible Systems

The interruptible systems are generally used to obtain information about the transmission medium. Is it transmitting light from the source to the detector or not? If it is not, the light beam has been interrupted. The interruptible systems have a common characteristic—the source and detector are separated to allow objects to interrupt the light beam. Systems of this type are used to open doors automatically, count products on a conveyor belt, determine liquid levels, determine speeds and positions of objects, and read barcodes and computer punched cards.

4. Optically Coupled Electronic Systems

Non-Interruptible Systems

The non-interruptible systems are used to pass information from the source to the detector with a transmission medium that is constant. Such systems can be used in a number of different ways. First they can be used simply to determine if the source is on or off. Next, they are used for systems that have some unique feature. For example, to provide electrical isolation between source and detector, to provide a communication link to a computer, or to provide a wide bandwidth information channel. For both the interruptible and non-interruptible system, similar considerations influence the design and construction of the systems. Some of these considerations are discussed in the next sections.

OPTICALLY COUPLED SYSTEMS DESIGN

Matching

A very important consideration in optically coupled systems is matching the spectral response of the light source and detector. The compatibility of source and detector can be determined by comparing the spectral response of the source and detector on the same graph as in *Figure 4-1*. If the wavelengths produced by the light source do not overlap the wavelengths to which the detector responds, the two are said to be mismatched. The detector could not "see" the light at all. It would be like transmitting a TV signal on channel 5 and expecting to receive it on channel 28. If there is some overlap as in *Figure 4-2*, the detector would have some response to the source but not as much as the pair shown in *Figure 4-3*.

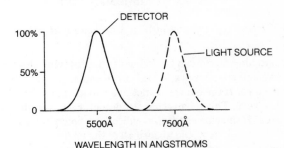

Figure 4-1. *Relative Response of Completely Mismatched Light Source and Detector*

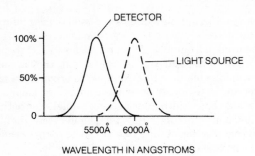

Figure 4-2. *Relative Response of Partially Matched Light Source and Detector*

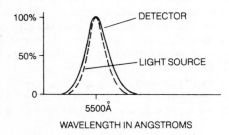

Figure 4-3. *Relative Response of Well Matched Light Source and Detector*

A matching factor can be defined by multiplying the detector response curve and the source spectral curve and measuring the area under the resulting curve. If the curves do not overlap at all, as shown in *Figure 4-1*, the area will be zero. This means that the detector will not absorb any energy from the source. If the curves totally overlap as shown in *Figure 4-3*, the area is unity, and the detector would absorb all of the energy from the source that falls on the detector. *Figure 4-2* is in between the two extremes of *Figure 4-1* and *4-3*. Here only a portion of the energy radiated by the source is absorbed by the detector.

Table 4-1 shows typical matching factors for a few common source-detector pairs. Two detectors are shown; the eye and a detector made from a silicon semiconductor junction. Examining these factors, it is noted that the silicon detectors are better detectors than the eye for all sources shown in *Table 4-1*.

4. OPTICALLY COUPLED ELECTRONIC SYSTEMS

Table 4-1. *Relative Matching Factors for Several Sources with the Eye and Silicon Junctions Compared as Detectors*

Source	Detector Matching Factor	
	Eye	Silicon PN Junction
Sun	0.16	0.5
2200°K Tungsten Lamp	0.007	0.19
2600°K Tungsten Lamp	0.021	0.24
3000°K Tungsten Lamp	0.044	0.3
Neon Lamp	0.35	0.7
Gallium Arsenide LED at 9000Å	0	1.0
Gallium Phosphide LED at 7000Å	0.08	0.7

If, for example, a system were to be designed to use a gallium arsenide LED at 9000Å as the source, *Table 4-1* shows that the eye would be a very poor detector; in fact, it would have no response at all. For such a system, the silicon detector is the logical choice. On the other hand, if the eye is the chosen detector, then the best possible source of the ones listed in *Table 4-1*, would be the neon lamp. The eye would absorb 35% of the energy that is radiated from the neon lamp source and strikes the eye. This doesn't mean that the neon lamp is the brightest, only that a larger percentage of the light wavelengths from the neon lamp are detectable by the human eye.

Selection of Transmission Medium

After assuring ourselves that the source and detector in the optical system are properly matched, we need to examine the other important component of the optical system—the medium through which information will be transmitted from the source to the detector. In both interruptible and non-interruptible systems, it is necessary at some point to have an uninterrupted transmission of light at some band of desired frequencies between the source and detector. Hence the medium must be able to pass light at the desired frequencies with minimal loss or attenuation. Additionally, in an interruptible system, we must be sure that the medium which will interrupt the light beam will cause the desired attenuation of the transmitted beam.

Optically Coupled Electronic Systems

In most applications for interruptible systems, the source and detector are separated by a distance such that other objects or materials fit between the source and detector to interrupt the radiated light. However, in uninterruptible systems, the source and detector may be very close together. In fact, they may be in the same unit as they are in the device called an optical isolator. For these devices, where the source and detector are separated only by small distances (10 to 50 mils), the electrical insulating properties of the medium become quite important for ensuring good electrical isolation.

Use of Air as a Medium

Figures 4-4 and *4-5* show examples of two interruptible systems which use air as the main transmission media. In *Figure 4-4*, the source and detector units are separated by an open space. Interruption of the light beam is accomplished by inserting a medium which will not pass the band of frequencies to which the detector is sensitive. In this case, the tabs on a revolving disc are the interrupters that block the light transmission between source and detector. Light current in the detector is turned from on to off as the transmission is interrupted. The fact that light current goes off and on indicates that the disc is in motion. The frequency or rate of the pulses indicate how fast the disc is revolving. The source in this case is a light-emitting diode and the detector is a semiconductor PN junction device.

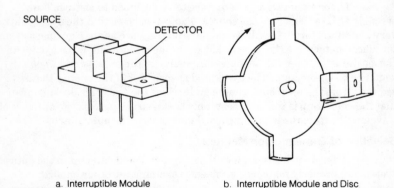

a. Interruptible Module   b. Interruptible Module and Disc

Figure 4-4. The Source-Detector Pair are Interrupted by Tabs on the Disk to Provide Information About Motion of the Disk

4
Optically Coupled Electronic Systems

In *Figure 4-5*, the arrangement of the source (a light emitter) and detector requires that the emitted light beam be reflected back to the detector in order to have an uninterrupted signal. In this case, the interruption of the signal is accomplished by inserting a medium into the transmission path which is not reflective at the frequencies to which the detector is sensitive. Again, the source is a light-emitting diode and the detector is a semiconductor PN junction device.

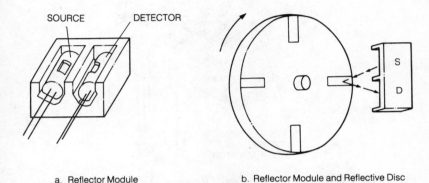

a. Reflector Module

b. Reflector Module and Reflective Disc

Figure 4-5. *The Detector Senses a Change in the Light Reflected from the Disk*

Use of an Electrical Insulating Medium

In *Figure 4-6*, we have a cutaway drawing which is an example of a non-interruptible optically coupled system. It is the optical isolator we mentioned previously. The source and detector are electrically isolated from each other and are packaged together in the device called an opto-coupler integrated circuit. The source and detector are placed physically close together inside the package and the space between them is filled with a medium which is transparent to the band of frequencies to which the detector is sensitive. Here is the case where the medium is selected to have good electrical insulating properties so that the electrical isolation is maintained even though the source and detector are very close together. The source in this case is a light-emitting diode and the detector is a phototransistor made from silicon material.

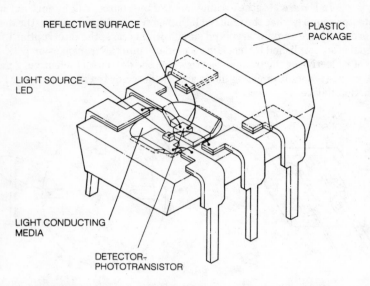

Figure 4-6. *Illustrates Construction of Photocoupled Source-Detector Pair*

Other Transmission Medium

A large variety of transmission media other than air is available for optically coupled systems. Various types of windows, optical filters, lenses, mirrors, prisms, diffraction gratings, and fiber optics may be used. Each may be used singly or in combination in any one system.

Windows

Windows are just what the name implies; a covering or closure that light can pass through from the packaged light source to the outside or from the outside to the packaged detector. They also, as shown in *Figure 4-6*, can be the transparent insulator between the physically close source and detector of the opto-isolators. Windows may be made of many types of materials including glass, plastic, quartz, and sapphire. Windows are designed to pass a wide range of light frequencies without attenuation.

4 OPTICALLY COUPLED ELECTRONIC SYSTEMS

Optical Filters

Optical filters are different. They are not made to pass all frequencies (wavelengths) of light energy. They are made to pass only selected bands of frequencies. For example, the usual functions of optical filters are to block out unwanted frequencies of light while letting desired frequencies pass through unattenuated. They also may be applied to uniformly attenuate a very broad band of frequencies of light energy. For example, a red optical filter passes only the frequencies of light which the eye detects as having the color red and severely attenuates all other light frequencies. Optical filters are usually made of very thin layers of metal deposited on a base made of glass or quartz.

Lens

A lens has characteristics very similar to a window in that it has to pass radiated light energy without attenuation; however, it is different in that it is used in optically coupled systems to focus or direct the light energy to particular points or in particular directions. It is used to direct a beam of light to cause it to converge (come to a point) or diverge (spread out). When a beam of light hits a lens which is transparent to the frequencies of light present in the beam, most of the light passes through but a small part of the light beam is reflected back to the source. One of the desirable qualities for lenses in an optically coupled system is that the reflected portion be as small as possible.

Figure 4-7 shows how a lens is used in the transmission path to concentrate the emitted light on the detector. *Figure 4-7a* shows a single lens being used and *Figure 4-7b* shows a double lens arrangement.

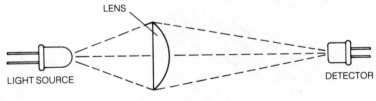

a. Single Lens Focusing System

b. Double Lens Focusing System

Figure 4-7. *Illustrates the Use of a Single Lens and Double Lenses for Focusing Light on Detector*

OPTICALLY COUPLED ELECTRONIC SYSTEMS

4

Figure 4-8 shows another application of lenses in a bar-code reader used to identify products in grocery or department stores. The bar code is a series of wide and thin lines separated by white space. Here the application of the lens is a bit different than in *Figure 4-7*. The light from the source (light-emitting diode) is focused by lens 2 to illuminate (light up) the bar code. Lens 1 focuses the reflected light from the bar code and magnifies it to produce an enlarged image for the photodiode detector.

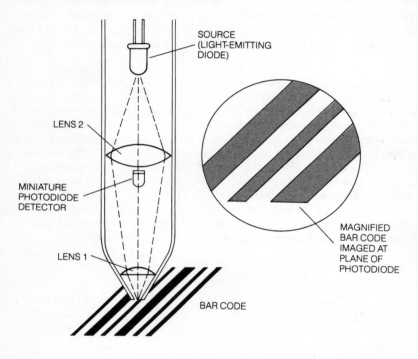

Figure 4-8. *Bar Code Reader Using an LED to Illuminate the Bar Code. The Reflected Signal is Detected by the Miniature Photodiode*

Since the transmission properties of a lens are adversely affected by any dust which collects on the lens, ways have been developed to help reduce the amount of dust reaching a lens. *Figure 4-9* shows an example of a dust protection tube which uses baffles to catch the dust before it reaches the lens and uses blasts of compressed air to force the dust away from the lens and back into the baffles.

4. Optically Coupled Electronic Systems

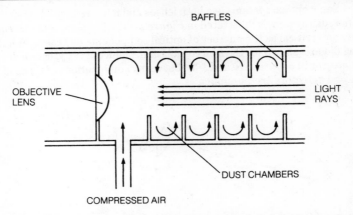

Figure 4-9. Dust Protection Tube with Baffles and Compressed Air

<u>Mirrors</u>

Mirrors reflect light and, in most applications, do not pass light. Mirrors are used to direct light beams and, like lenses, cause the light beams to converge and diverge. When a beam of light hits a mirror, most of the light is reflected but a small portion of the light beam is transmitted into the mirror instead of being reflected. A desirable quality for a totally reflecting mirror in an optically coupled system is that it reflect very close to 100% of the light which hits it.

Figure 4-10 shows the use of a curved mirror to reflect the emitted light from the source back to the detector. The shape of the mirror and the relative distances between the mirror and detector and between the mirror and source focus the reflected light into a concentrated beam at the detector. Alignment of the mirror may require careful adjustment to ensure proper operation.

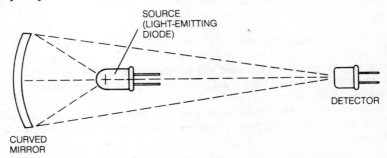

Figure 4-10. Application of Curved Mirror to Reflect and Focus Light on Detector

OPTICALLY COUPLED ELECTRONIC SYSTEMS

4

Some optical systems use both lenses and mirrors in the transmission path. One such system is illustrated in *Figure 4-11*. Alignment and focusing become more critical as the number of optical components increases. A light-emitting diode is shown for the source and semiconductor PN junction device as the detector in *Figure 4-10* and *4-11* but many combinations of types of sources and detectors are possible.

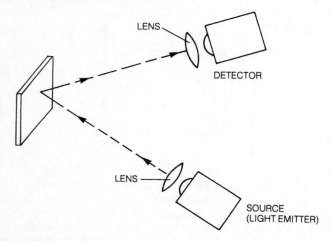

Figure 4-11. *Application of Both Lenses and Mirror to Reflect and Focus Light on Detector*

<u>Prisms and Diffraction Gratings</u>

Prisms and diffraction gratings may also be used to process light beams. These devices can be used to separate a beam of light into its spectral components. Several detectors could be used to provide information about the wavelengths contained in the light provided by the source.

<u>Prism</u>

A prism works because the velocity of light in glass depends on the wavelength. As a result, a beam of light containing two wavelengths and striking the prism at the same point will exit the prism at two different points as shown in *Figure 4-12a*. *Figure 4-12b* shows a possible application which would allow the detectors to provide information to determine the status of two different color lights. Detectors located at a remote location could be used to sound an alarm if the red light is on or provide a control signal if the green light is on.

4. Optically Coupled Electronic Systems

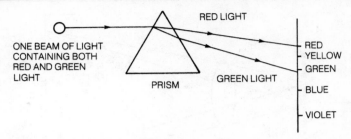

a. Prism separates a beam of light into its components by wavelength (color)

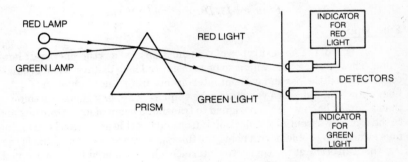

b. Application of a prism

***Figure 4-12.** Prism*

Diffraction Grating

The diffraction grating shown in *Figure 4-13* is made of an opaque material and produces interference patterns when light travels through the slits. As a wave front of light strikes the opaque material, each slit changes the velocity of light close to the edges of the slit more than at the center of the slit. The amount of change depends on wavelength. If there are several slits, the resulting wavefronts will reinforce each other at some points (bright light) and cancel at other points (black). The image on the screen (*Figure 4-13*) will have bars of light of different colors separated by black bars.

When comparing the prism to the diffraction grating, it should be noted that the spectral lines produced by the prism get closer together as the wavelengths get longer. The diffraction grating, on the other hand, produces spectral lines that are uniformly spaced. This may make it easier to detect differences in the spectral lines at longer wavelengths by using the diffraction grating. At shorter wavelengths, it may be better to use a prism. It should be noted also that the defraction grating produces two patterns on each side of zero where the prism produces only one.

OPTICALLY COUPLED ELECTRONIC SYSTEMS

4

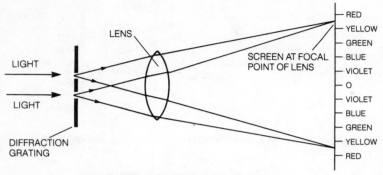

Figure 4-13. Diffraction Grating

Fiber Optics

One of the newest types of transmission media is the fiber optic light guide. These devices allow light to be transmitted long distances and around corners with little loss and without interference from other light sources. A fiber optic light guide is a very thin tube of quartz or epoxy (some are thinner than a human hair) which is designed to transmit a beam of light from one end to the other by essentially reflecting it from side to side as it travels down the fiber. The main problem with using the fiber as a transmission medium is the loss that can occur at the connector interface where the light beam is coupled into and out of the fiber. However, once the beam is in the fiber, it is transmitted very efficiently from one end to the other. Some typical fibers are shown in *Figure 4-14* and some of the factors affecting loss are illustrated and compared in *Figure 4-15*. (Note that the term "flux" is used to indicate the power of the light in watts or candelas.) A large loss occurs when coupling the source to the fiber, a smaller loss occurs when coupling the fiber to the detector and a small loss per unit of length occurs down the length of the fiber from the source to detector.

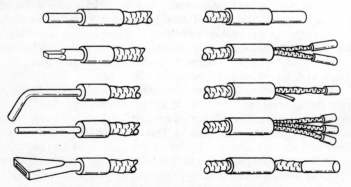

Figure 4-14. Typical Fiber Optic Geometries

4. OPTICALLY COUPLED ELECTRONIC SYSTEMS

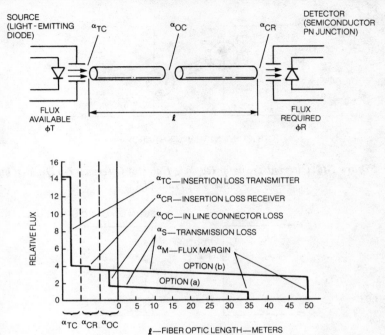

Figure 4-15. Shows Loss of Signal Strength (Flux) in Fiber Optic System

WHAT ARE THE IMPORTANT CHARACTERISTICS OF AN OPTICALLY COUPLED SYSTEM?

There are four main characteristics of optically coupled systems. They are the current transfer ratio (CTR) or gain, the frequency response, the amount of electrical isolation, and the immunity to both internal and external noise sources.

Transfer Characteristics

If the optically coupled system is being used in an analog (linear system) application, the CTR is the ratio of the output current from the detector to the input current into the source. This is shown in *Figure 4-16* for an optoelectronic device consisting of a source and detector. The CTR is not constant but varies as a function of the input current. As shown in *Figure 4-17*, it also depends on the transmission medium.

OPTICALLY COUPLED ELECTRONIC SYSTEMS

4

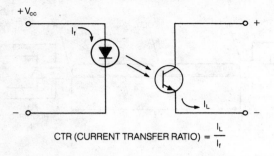

Figure 4-16. *Current Transfer Ratio is Detector Current (I_L) Divided by Source Current (I_f)*

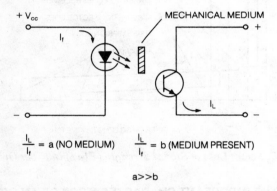

Figure 4-17. *The Transfer Ratio is Reduced by Partially Blocking the Light*

If the optically coupled system is being used in a digital application, the CTR may be given differently. In digital binary systems there are only two voltages levels, high and low. The output voltage is either one or the other depending on the value of the input signal. As a result the input versus output function includes a current to voltage transfer and is something like *Figure 4-18*. It can be either like curve 1 or curve 2.

As the input is increased from zero, no change occurs in the output until a threshold point (#1) is reached. Increasing the input beyond the threshold causes the output to change rapidly to a new level. Further increases in the input causes little change in the output. Throughout the digital system the levels will either be at level A or level B. As a result the input current to a light source for an optically coupled system must be related to level A or level B.

4
OPTICALLY COUPLED ELECTRONIC SYSTEMS

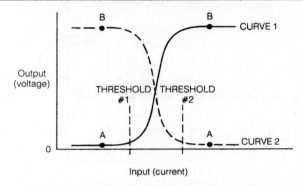

Figure 4-18. Typical Output vs Input Function for Digital System

One way of specifying CTR in digital systems is to consider that the horizontal axis of *Figure 4-18* is the input current to the source and that the vertical axis is the output voltage of the detector circuit. For this example, let's choose curve 2. The CTR is specified by saying that for a source current greater than or equal to a threshold value (#2), the output voltage will be less than or equal to level A.

As an example, consider a photocoupled system with the detector driving a standard SN7400 series TTL logic gate. Suppose that point B on curve 1 of *Figure 4-18* is at an input current to the source that produces a detector output current of $I = 1.6mA$ in *Figure 4-19a*. The current I turns on T1 and turns off T2 and the output voltage level of the gate is at point B. If now the input current to the source is reduced to zero, the detector current I is only the small value of dark current (T1 is off, T2 is on) and the output voltage of the gate is at the low logic level (point A). If the source current is present, it must be large enough to generate enough light so that the detector can sink (draw from the logic gate) at least 1.6mA in order for the gate output to be high (point B). When the source current is zero, the detector dark current should be small enough so that T1 does not turn on.

If four logic gates are to be driven by the detector (*Figure 4-19b*), then the detector must be able to sink at least four times the current required for one gate ($4 \times 1.6mA = 6.4mA$). If the detector is capable of sinking 6.4mA, the circuit will work reliably.

The number of circuits that can be driven is called a circuit "fan-out". Therefore, the CTR of an optically-coupled system's source and detector pair may be specified as the logic circuit fan-out for a given input current to the source. The fan-out of the current of *Figure 4-19b* is four. The higher the fan-out for a given input source current, the higher the CTR.

Optically Coupled Electronic Systems

4

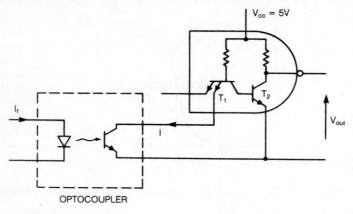

a. Optically Coupled System and Simple Equivalent Circuit of TTL logic Gate

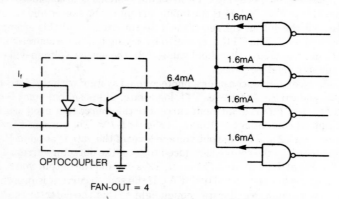

b. Fan-out of an Optically Coupled System Driving Digital IC's

Figure 4-19. Fan-out of a Digital Circuit

Frequency Response

The second important characteristic of optically coupled systems is frequency response. *This characteristic must not be confused with the spectral response of the system.* The *spectral response* is the range of light frequencies which will cause a detectable current to flow in the detector. The *frequency response* of the optically coupled system is a measure of how rapidly the system can respond to changes in the input signal such as switching from off to on. If the intensity of the light beam is varied (modulated) by some signal, the spectrum of the beam stays constant but the system must be able to change the output current to follow the corresponding changes in intensity.

4 OPTICALLY COUPLED ELECTRONIC SYSTEMS

Switching Times

One method of determining the frequency response for optically coupled systems is to measure switching times. To do this, a current pulse is generated at the source and the corresponding output current pulse produced at the detector is measured. The various switching times that are measured are shown in *Figure 4-20*. The delay time t_d is the time from turn-on of the input pulse until the output pulse amplitude rises to 10% of its fully-on value. Rise time, t_r, is the time for the output pulse amplitude to rise from 10% to 90% of its fully-on value. Turn-on time t_{on} is the sum of t_d and t_r. The storage time, t_s, is the time from turn-off of the input pulse until the output pulse amplitude falls to 90% of its fully-on value. Fall time, t_f, is the time for the output pulse amplitude to fall from 90% to 10% of its fully-on value. Turn off time t_{off} is the sum of t_s and t_f. Pulse width, t_w, is the time between 50% points on the rising and falling portions of the output pulse.

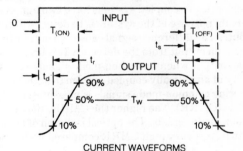

Figure 4-20. Switching Times

Frequency Limit (f_c)

Another way of characterizing frequency response for optically coupled systems is to specify a cutoff frequency of the system. This is the frequency at which the detector sensitivity is 0.707 of the value within a frequency range where the sensitivity is a maximum. The frequency limit is related to the rise time by the equation,

$$f_c = \frac{0.35}{t_r}$$

The equation can be rearranged as

$$t_r = \frac{0.35}{f_c}$$

to determine rise time. Thus, an optically coupled system with $f_c = 100$ kHz has a rise time of

$$t_r = \frac{100 \text{kHz}}{0.35} = 3.5 \mu s,$$

OPTICALLY COUPLED ELECTRONIC SYSTEMS

4

and a system with $t_r = 0.1\mu s$ has a limiting frequency of

$$f_c = \frac{0.1\mu s}{0.35} = 3.5 mHz.$$

Isolation

A third important feature of optically coupled systems is the property of isolation. The term isolation really includes four different properties; common-mode rejection, ground loop prevention, electrical insulation and electrical noise isolation.

<u>Common Mode Rejection</u>

Figure 4-21 shows the input circuit to an optically coupled system. Two types of signals are shown. A common mode signal is where the exact same signal appears on both inputs at the same time. A difference signal is where the difference between the signals on both inputs is the important information. When optically-coupled systems have two inputs as shown in *Figure 4-21*, the main purpose of the system is to detect a difference in the signals on the inputs. The difference signal causes an output from the source which in turn causes an output from the detector. It is undesirable for the detector to have an output when a common-mode signal excites the inputs.

The ability of an optically coupled system to detect difference mode signals while rejecting common-mode signals is measured by the common mode rejection ratio (CMRR). The higher the CMRR, the better the system is at rejecting common mode signals. The optical isolation of an optically coupled system causes it to have a very high CMRR compared with other types of coupling.

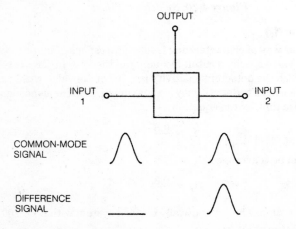

Figure 4-21. Types of Signals on Inputs of Optically Coupled Systems

4 OPTICALLY COUPLED ELECTRONIC SYSTEMS

Ground Loop Prevention

Another isolation property involves the ability of the transmission medium to resist the flow of electrical current through it. This helps to prevent the flow of undesirable ground loop currents. As shown in *Figure 4-22*, a ground loop current may occur in a system when several different grounds are used at different locations in the circuit. Ideally, all ground points should be at the same potential, but if a current can find a path to flow between the various grounding points, undesirable voltage drops are generated between these ground points. These voltage drops upset the normal operation of the circuit. The use of optically isolated systems can be very helpful in eliminating these ground loops because electrical isolation is provided by the light transmission path between source and detector. As a result ground loop paths are not completed and ground loop currents do not flow.

Electrical Insulation

A third isolation property involves the resistance of the transmission medium to electrical breakdown when a relatively high voltage is applied across it. As discussed previously, this is a reason why the transparent materials used between the source and detector in opto-isolators are chosen with regard to their properties as electrical insulators as well as their ability to pass the desired frequencies.

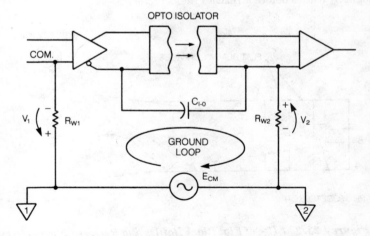

Figure 4-22. *Ground Loop Current Through C_{I-0} Sets Up Undesirable Voltage Drops V_1 and V_2. Opto-isolator Reduces the Effect of These Ground Loop Voltages*

OPTICALLY COUPLED ELECTRONIC SYSTEMS

4

Electrical Noise Isolation

Because the source and detector are separated by a light transmission path, electrical noise generated in each of the individual electrical circuits is isolated from the other. This is especially important in keeping electrical noise generated in the source circuit from reaching the detector.

Noise Immunity

A final major characteristic to be considered is the ability of the optically coupled system to resist interference from external and internal noise sources. This property is called noise immunity.

External Noise

External noise means unwanted signals. It consists of light radiated from some source other than the desired source; that is, at frequencies to which the detector is sensitive. Some of the ways to reduce the effects of external noise are to: (1) ensure that the desired source light signal is more powerful than the external noise, (2) change the operating frequencies of the optically coupled system so the detector does not respond to the external noise, and (3) shield the optically coupled system with an enclosure that will attenuate the external noise before it reaches the detector. Such an enclosure is shown in *Figure 4-23*. Ambient light that could cause external noise is deflected and caught by baffles before it reaches the detector.

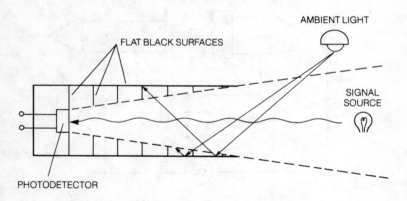

Figure 4-23. *The Use of Flat Black Baffles and Recessed Photodetector to Minimize the Effect of Ambient Light*

4-22   UNDERSTANDING OPTRONICS

4 — Optically Coupled Electronic Systems

Internal Noise

The effect of internal noise in an optically-coupled system is to cause an output current to flow in the detector when there is no signal from the source. There are three main sources of internal noise in the detector—thermal noise, shot noise, and contact noise. Thermal noise (also called Johnson noise and Nyquist noise) is due to the thermal agitation of electrons (current flow caused by heat) within the resistive elements of the detector. Shot noise is caused by a current flow across a potential barrier in semiconductor materials. Contact noise, also called flicker noise or 1/f noise because of its dependence on the frequency f, is due to an imperfect contact between two materials. Contact noise can occur in a transistor or diode where different metals are bonded together, or a carbon resistor made up of many small particles molded together. The output currents due to these three noise sources add together to make up the total internal noise current of the detector.

Signal-to-Noise Ratio

A measure of an optically coupled system's immunity to internal noise is given by the signal-to-noise ratio. This is defined as the ratio of the output current when a signal is applied to the output noise current present due to internal noise when no signal is being applied.

Another measure of noise immunity derived from using the signal-to-noise ratio is the noise equivalent power, NEP. This number is essentially the input signal flux level that would cause a signal-to-noise ratio of one. In other words, NEP is the amount of input signal flux to the source necessary to produce an output current in the detector which is equal to the noise current.

WHAT HAVE WE LEARNED?

1) An optically coupled electronic system is a system which uses light from a source to pass information directly or indirectly to the detector circuit.
2) Optically coupled systems fall into two categories—interruptible and non-interruptible.
3) Optically coupled systems consist of a source, a detector, and a transmission medium.
4) Control of the source, detector, and medium are necessary to provide a reliable optically coupled system.
5) Important characteristics of an optically coupled system are current transfer ratio, frequency response, isolation, and noise immunity.

OPTICALLY COUPLED ELECTRONIC SYSTEMS 4

Quiz for Chapter 4

1. Light provides information when there is
   a. information in the light.
   b. a detector which will respond to the light.
   c. a system to interpret the signals from the light.
   d. all of the above.

2. An optically coupled system can be designed to
   a. detect motion.
   b. read printed information.
   c. carry information from a computer to a printer.
   d. all of the above.

3. Important characteristics of an optically coupled system are the
   a. matching factor of the source detector pair.
   b. content of information.
   c. frequency response.
   d. none of the above.

4. Lenses can be used to control the
   a. source.
   b. detector.
   c. transmission medium.
   d. none of the above.

5. Light fibers are used to
   a. emit light.
   b. sense light.
   c. guide light.
   d. all of the above.

6. CTR is an important characteristic of the
   a. media.
   b. detector.
   c. source.
   d. system.

7. An incandescent lamp would not be as well suited as an LED to transfer information through a light fiber because of its
   a. intensity.
   b. wavelength.
   c. frequency response.
   d. none of the above.

8. Signal-to-noise ratio is a measure of
   a. CTR.
   b. the ability of the system to resist interference from undesired internal sources.
   c. none of the above.

9. *Figure 4-4* is a
   a. reflective reader.
   b. photo coupled isolator.
   c. interruptible system.
   d. non-interruptible system.

10. The source detector pair in *Figure 4-1* would
    a. not be compatible.
    b. not be efficient.
    c. be a good choice for an optically-coupled system.
    d. none of the above.

1. d, 2. d, 3. a, 4. c, 5. c, 6. d, 7. c, 8. b, 9. c, 10. a

5 OPTOELECTRONIC DISPLAYS

Optoelectronic Displays

ABOUT THIS CHAPTER

Consider for a moment a rule that seems to make sense: things are made for use only when there is a need for them (or, to use a well worn cliche; necessity is the mother of invention). The development of displays was spurred by a need. What was that need? How has the need been affected by time?

Webster defines display as "a device that gives information in visual form in communications". A display may be a billboard, a picture, a printed page, fireworks in the sky, a sign, flags, or any of the other means of attracting attention and conveying information. These are common everyday interpretations of the word display. For our purposes in this chapter, we will be talking about how electronics, particularly optoelectronics, is being used to form "a display" to visually communicate information to meet a need.

WHY IS A DISPLAY NEEDED?
Person-to-Person Communication

The need for a display is fundamental for visual communication. Without a display, we are limited to oral communication to convey concepts and facts from one person to another. Even a few lines drawn in the sand form a simple display that may aid in communication.

Significant advances have been made in visual communications in the past. The advent of the printing press made it possible for one person to communicate to many people by means of visual communication, even people at remote locations. The development of photography provided another means of visual communication and also improved printing because the application of photographic techniques has improved the quality and speed of printing. Further development of these photographic techniques produced moving pictures which provide visual communication of amazing detail and realism.

One disadvantage of communication by photographs and the printed page was that they had to be physically carried to another location. For this reason, facsimile was developed to convert pictures and printed material to electrical signals so they could be rapidly transmitted over wires to a remote location for exact reproduction on paper. Although facsimile is still quite useful, it has been replaced in some applications by high-speed data communication and television. Television was developed primarily by combining the technologies of radio and a display technology developed for radar using the cathode ray tube (CRT). This is perhaps today's fastest and most effective means of bringing both visual and audible information from a few to many.

OPTOELECTRONIC DISPLAYS 5

Machine-to-Person Communication

Until recent history, communication was from person-to-person, but now communication is often between people and machines. People have developed many machines to serve their needs and pleasure, and in many cases, the purpose of the machines can be carried out only if visual information is provided by means of a display. A television receiver, for example, would be of little use if the picture tube did not produce a visible display of the information being transmitted.

Another example is the computer. With the information explosion, it has become increasingly necessary for people to put information into a computer for manipulation and storage. The stored information may be recalled for display on a local CRT or printer, or transmitted rapidly to many people all over the world.

Various types of displays are used on microwave ovens, washing machines, automatic bank teller machines, watches, and electronic games to indicate current status, provide instructions, or display other information.

Displays such as the CRT can extend the capabilities of humans. An example is the use of a television camera and receiver to extend human eyes to remote or unsafe locations. The display may be used simply for observation or inspection, or robot arms may be controlled to perform some function by using the display as a guide. The driving force behind all these developments is a desire to see what normally cannot be seen, or to do those things that normally have not been done.

Many optoelectronic displays, however, are far less complex in function and construction than television. Such displays are used to simply communicate the status of something. Applications are common in automobiles, aircraft, machinery, and test equipment. Examples are warning indicators for power on or off, low oil pressure, high temperature, park-brake on, table position, in calibration, etc. For these uses, only a single simple lamp is required for the display.

So we see that we need displays for both person-to-person communication and machine-to-person communication. The display type can range from a single lamp for simple tasks to the CRT for large volumes of complex information.

WHAT ARE THE DISPLAY TYPES?

There are several ways one might choose to classify displays, including optoelectronic displays. The following sections classify them based upon their functional roles in order of their functional complexity. These categories are indicators, numeric displays, alphanumeric displays, and special function displays.

5 OPTOELECTRONIC DISPLAYS

An alternative classification would have been by device technology. Technologies have already been discussed in general for semiconductor devices that are a source of light (light emitters). Displays which consist of light-emitting or light-path control elements may be designed using any of the device technologies discussed.

Indicators

An indicator is normally a single lamp or VLED and represents the least complex display. Very little information can be gained from a single lamp; it is either on or off. Yes, it can also be dim or bright, but this is of little use unless there is a reference level, and even then the eye can barely detect a 2:1 brightness-contrast ratio. However, indicators can be used to communicate more complex information by innovative techniques as illustrated by Paul Revere's warning, "One if by land, two if by sea". Ships have long used a single light source with shutters that open and close to send messages by means of an on-off light code.

More complex information can be represented by using the individual light sources in combinations according to a code. Linear arrays of lamps may be used to display binary (base 2) numbers. *Table 5-1* shows how a 4-element linear array can display 16 different pieces of information which, with a code, can represent decimal (base 10) numbers from 0 through 15. A similar 8-element linear array can display 256 different pieces of information. The disadvantage of displaying information in this format is that most people do not easily recognize the binary number system and find it difficult to read binary information. Displays have been developed; therefore, to display numbers in the decimal system that people are accustomed to using.

Table 5-1. Use of Indicator Lamps to Display Numeric Information

Indicator *	Base 10 Number	Base 2 Number
○ ○ ○ ○	0	0 0 0 0
○ ○ ○ ●	1	0 0 0 1
○ ○ ● ○	2	0 0 1 0
○ ○ ● ●	3	0 0 1 1
○ ● ○ ○	4	0 1 0 0
○ ● ○ ●	5	0 1 0 1
○ ● ● ○	6	0 1 1 0
○ ● ● ●	7	0 1 1 1
● ○ ○ ○	8	1 0 0 0
● ○ ○ ●	9	1 0 0 1
● ○ ● ○	10	1 0 1 0
● ○ ● ●	11	1 0 1 1
● ● ○ ○	12	1 1 0 0
● ● ○ ●	13	1 1 0 1
● ● ● ○	14	1 1 1 0
● ● ● ●	15	1 1 1 1

* ○ Represents an indicator which is off (binary zero).
● Represents an indicator which is on (binary one).

OPTOELECTRONIC DISPLAYS

5

Semiconductor Numeric Displays

One type of numeric display for decimal numbers consists of arrays of discrete (individually packaged) indicators (*Figure 5-1a*) arranged to display the numerals 0 through 9. Another type of numeric display which is commonly used is the seven-segment display as shown in *Figure 5-1b*. Each of the segments is a separate VLED. By turning on only certain segments, any of the numerals 0 through 9 can be displayed. Actually 2^7 or 128 different displays are possible with seven segments, but only the combinations shown at the bottom of *Figure 5-2* are used. These combinations can be displayed with only a 4-bit input (4 on-off control lines). Notice that this arrangement uses seven indicators to display the numbers 0 through 9 where the binary code array in *Table 5-1* requires only four indicators to represent the same numbers. Efficiency in the use of electrical power is sacrificed to gain efficiency in the usability of the display.

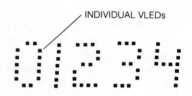

a. Display Formed By Individual VLEDs

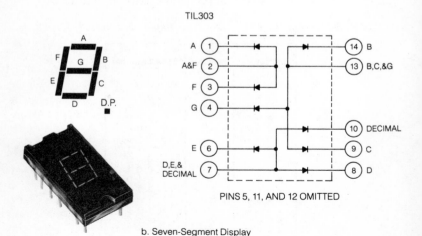

b. Seven-Segment Display

Figure 5-1. *Numeric Displays*

5 OPTOELECTRONIC DISPLAYS

Decoder/Driver for Displays

The fundamental purpose of VLED displays of this type is to display the condition of a set of outputs from a computer or microprocessor that has a binary coded signal on the outputs. In order to do this, the binary coded outputs must be decoded and driver circuits must be used to supply the current required by the display. *Figure 5-2* shows the schematic diagram of how a typical seven-segment optoelectronic display is packaged and gives typical application data for such a display. The segments consist of two light-emitting diodes in series which are driven by a transistor which is the output stage of a decoder/driver integrated circuit (SN7447A). This IC decodes the input binary code and energizes the driver transistor so that the correct segments are lighted to display the decimal number represented by the code.

The SN7447A has 4 inputs A, B, C, and D which accept the binary coded signal from a source such as a computer memory. The binary code may be any one of the codes shown in *Table 5-1*. The 0 and 1 indicate a voltage level into the inputs A, B, C, and D. 0 is a low voltage and 1 is a high voltage. The construction of the seven-segment display consists of segments A through G and the decimal point D.P., which light up depending on the input code on A, B, C, and D. Look at the first line in the table of *Figure 5-2*. Notice that when inputs A, B, C, and D are low, the segment lines a, b, c, d, e, and f are ON while g is OFF. This displays a numeral zero because the binary number 0 0 0 0 is equal to the decimal number 0.

Now look at the seventh line in the table. Input lines C and B are high while D and A are low to represent binary number 0 1 1 0. The lighted segments are c, d, e, f, and g which displays the numeral 6 as shown in *Figure 5-3*. Referring to *Table 5-1* shows that the base 2 number 0 1 1 0 is equal to 6 in base 10.

The decoder/driver circuit in this case has two control inputs BI/RBO (blanking input/ripple blanking output) and RBI (ripple blanking input). The column in the function table labeled BI/RBO indicates the voltage level on the control line. The BI/RBO line must be high in order for the coded input on input lines A, B, C, and D to be active. If BI/RBO is low, all segments will be turned off and the coding will be ineffective. By controlling the voltage on BI/RBO, the complete display unit can be turned on or off. The RBI controls a special feature which allows the zero in front of a decimal point to be suppressed.

To conserve power in displays such as this, the display can be pulsed on and off (scanned) by use of BI/RBO functions. This also may allow a computer or microprocessor to control more than one digit on a time-shared basis. This is called multiplexing and allows a computer to provide information to several digits from only one set of data lines. The human eye averages the light output of the displays and never notices that the displays are actually flashing on and off.

OPTOELECTRONIC DISPLAYS 5

TYPICAL APPLICATION DATA

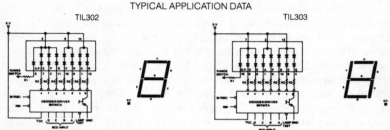

NOTES: A. R1 and R2 are selected for desired brightness.
B. SN74L47 may be used in place of SN7447A in applications where segment forward current will not exceed 20 mA.

FUNCTION TABLE SN7447A

DECIMAL OR FUNCTION	INPUTS					BI/RBO†	SEGMENTS							NOTE
	LT	RBI	D	C	B	A		a	b	c	d	e	f	g	
0	H	H	L	L	L	L	H	ON	ON	ON	ON	ON	ON	OFF	1
1	H	X	L	L	L	H	H	OFF	ON	ON	OFF	OFF	OFF	OFF	1
2	H	X	L	L	H	L	H	ON	ON	OFF	ON	ON	OFF	ON	1
3	H	X	L	L	H	H	H	ON	ON	ON	ON	OFF	OFF	ON	1
4	H	X	L	H	L	L	H	OFF	ON	ON	OFF	OFF	ON	ON	1
5	H	X	L	H	L	H	H	ON	OFF	ON	ON	OFF	ON	ON	1
6	H	X	L	H	H	L	H	OFF	OFF	ON	ON	ON	ON	ON	1
7	H	X	L	H	H	H	H	ON	ON	ON	OFF	OFF	OFF	OFF	1
8	H	X	H	L	L	L	H	ON	ON	ON	ON	ON	ON	ON	1
9	H	X	H	L	L	H	H	ON	ON	ON	OFF	OFF	ON	ON	1
10	H	X	H	L	H	L	H	OFF	OFF	OFF	ON	ON	OFF	ON	1
11	H	X	H	L	H	H	H	OFF	OFF	ON	ON	OFF	OFF	ON	1
12	H	X	H	H	L	L	H	OFF	ON	OFF	OFF	OFF	ON	ON	1
13	H	X	H	H	L	H	H	ON	OFF	OFF	ON	OFF	ON	ON	1
14	H	X	H	H	H	L	H	OFF	OFF	OFF	ON	ON	ON	ON	1
15	H	X	H	H	H	H	H	OFF	OFF	OFF	OFF	OFF	OFF	OFF	1
BI	X	X	X	X	X	X	L	OFF	OFF	OFF	OFF	OFF	OFF	OFF	2
RBI	H	L	L	L	L	L	L	OFF	OFF	OFF	OFF	OFF	OFF	OFF	3
LT	L	X	X	X	X	X	H	ON	ON	ON	ON	ON	ON	ON	4

H = high level (logic 1 in positive logic), L = low level (logic 0 in positive logic), X = irrelevant.

†BI/RBO is wire-AND logic serving as blanking input (BI) and/or ripple-blanking output (RBO).

NOTES: 1. The blanking input (BI) must be open or held at a high logic level when output functions 0 through 15 are desired. The ripple-blanking input (RBI) must be open or high if blanking of a decimal zero is not desired.
2. When a low logic level is applied directly to the blanking input (BI), all segment outputs are off regardless of any other input.
3. When the ripple-blanking input (RBI) and inputs A, B, C, and D are at a low logic level with the lamp test input high, all segment outputs are off and the ripple-blanking output (RBO) of the decoder goes to a low level (response condition).
4. When the blanking input/ripple blanking output (BI/RBO) is open or held high and a low is applied to the lamp-test input, all segments are illuminated.

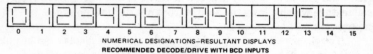

NUMERICAL DESIGNATIONS—RESULTANT DISPLAYS
RECOMMENDED DECODE/DRIVE WITH BCD INPUTS

***Figure 5-2.** Typical Application of Seven-Segment Display*

5 OPTOELECTRONIC DISPLAYS

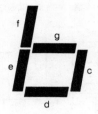

Figure 5-3. *Typical Display of Numeral 6*

Figure 5-4 gives a typical application of a seven-segment optoelectronic numeric display element used to display a three-digit number with sign for a counter. The function of the counter is to increment the display by one each time the counter input is cycled from low to high and back. The counter signal activates the A input of the SN7490 binary counter (IC_1). The outputs of this counter are fed to the SN7447A (IC_2) decoder/driver which causes the display (IC_3) to display the least significant digit. As soon as the count reaches ten, the D output of IC_1 provides a pulse to the A input of the next counter (IC_4). The outputs of IC_4 provide data to the decoder/driver (IC_5) which drives the display (IC_6). This process continues to the next digit IC_7, IC_8, and IC_9 as the count increases.

The pulsing output of an interrupted optically-coupled system can be applied to the A input of IC_1 to count the pulses that represent the number of objects on a conveyor belt that are interrupting the light beam. The seven-segment displays will show the increasing count as the objects pass and the final count when all of the objects have passed.

Fully Integrated

There are specialized displays with logic which have been developed for use as counters which also incorporate the decode and driver functions described in *Figure 5-4*. All of the circuitry is on an integrated circuit chip in one package. Because the package contains the necessary electronics for decoding, counting, and driving the displays; it requires very little external circuitry which, of course, simplifies its use in an application. *Figure 5-5* shows the required external connections. Comparison of these to the interconnections required for the system of *Figure 5-4* demonstrates how systems are becoming much more compact, lightweight, and reliable because of integrated circuit technology. The reduced number of packages and interconnections is a prime contributing factor to the increased reliability.

OPTOELECTRONIC DISPLAYS

TYPICAL APPLICATION DATA

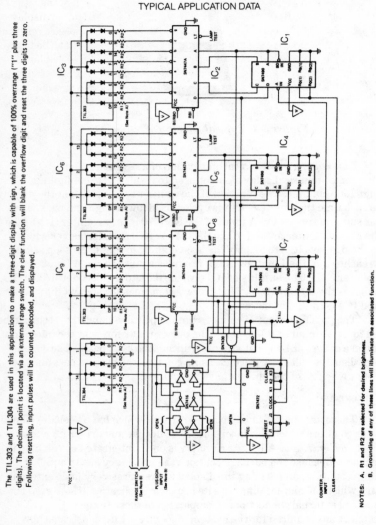

Figure 5-4. *Typical Application of Numeric Displays as a Counter*

5 OPTOELECTRONIC DISPLAYS

The latch strobe input for the circuit of *Figure 5-5* allows the counter to update while the display maintains the preceding count. When the latch strobe signal is activated, the count at that instant is latched in the display although the actual count may be changing. The display is updated every time the latch strobe signal is applied. By using the latch strobe signal, the blur of rapidly changing digits is eliminated and only a stable, readable display is presented. The overriding blanking input allows the display to be intensity modulated (dimmed) or turned completely off to conserve power while the counter circuits are still operating. The ripple blanking input allows the leading zeroes of the display to be blanked. These features give the designer a great deal of flexibility in using these counters.

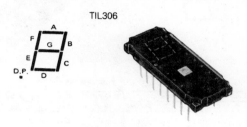

TYPICAL APPLICATION DATA

This application demonstrates how the displays may be cascaded for N-bit display applications. It features:

Synchronous, look-ahead counting
Ripple blanking for leading zeros
Overriding blanking for total suppression or intensity modulation of display
Direct parallel clear
Latch strobe permits counter to acquire data for the next display while viewing current display.

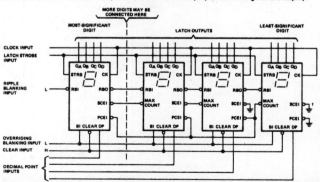

†The serial carry input of the least-significant digit is normally grounded; however, it may be used as a count-enable control for the entire counter (high to disable, low to count) provided the logic level on this pin is not changed while the clock line is low or false counting may result.

***Figure 5-5.** Numeric Displays with Onboard Counter*

OPTOELECTRONIC DISPLAYS

5

Numeric displays for use with binary inputs have also been developed to combine the decoder/driver function and the display into one package. A typical example and its function table is given in *Figure 5-6*. This is still a numeric unit even though several letters are included in the table. Compact units such as this make it easier for the system designer to arrive at a final system and save time and labor in assemblying systems.

TIL308

FUNCTION TABLE

FUNCTION	LATCH INPUTS						BLANKING INPUT	LED TEST	LATCH OUTPUTS					DISPLAY TIL308
	D	C	B	A	DP	STROBE			Q_D	Q_C	Q_B	Q_A	Q_{DP}	
0	L	L	L	L	L	L	H	H	L	L	L	L	L	0
1	L	L	L	H	H	L	H	H	L	L	L	H	H	.1
2	L	L	H	L	L	L	H	H	L	L	H	L	L	2
3	L	L	H	H	H	L	H	H	L	L	H	H	H	.3
4	L	H	L	L	L	L	H	H	L	H	L	L	L	4
5	L	H	L	H	H	L	H	H	L	H	L	H	H	.5
6	L	H	H	L	L	L	H	H	L	H	H	L	L	6
7	L	H	H	H	H	L	H	H	L	H	H	H	H	.7
8	H	L	L	L	L	L	H	H	H	L	L	L	L	8
9	H	L	L	H	H	L	H	H	H	L	L	H	H	.9
A	H	L	H	L	L	L	H	H	H	L	H	L	L	A
MINUS SIGN	H	L	H	H	H	L	H	H	H	L	H	H	H	-
C	H	H	L	L	L	L	H	H	H	H	L	L	L	C
BLANK	H	H	L	H	H	L	H	H	H	H	L	H	H	.
E	H	H	H	L	L	L	H	H	H	H	H	L	L	E
F	H	H	H	H	H	L	H	H	H	H	H	H	H	.F
BLANK	X	X	X	X	X	X	L	H	X	X	X	X	X	
LED TEST	X	X	X	X	X	X	X	L	X	X	X	X	X	.8

H = high level, L = low level, X = irrelevant.
DP input has arbitrarily been shown activated (high) on every other line of the table.

***Figure 5-6.** Numeric Displays with Logic*

5 OPTOELECTRONIC DISPLAYS

Semiconductor Alphanumeric Displays

With the increasing capabilities of computers came the recognition that words can be manipulated, stored, and retrieved as well as numbers. In fact, much of today's computer data banks is dedicated to storage of alphanumeric data. Alphanumeric data consists of letters, numbers, and symbols associated with text rather than for solving equations. Thus, we can see that a display is needed that is capable of displaying letters and the common symbols, as well as numbers.

A typical unit is shown in *Figure 5-7*. Notice that the arrangement of the individual light-emitting sources is different from that of the seven-segment display; it is a 5 × 7 matrix. The display consists of 5 columns and 7 rows; therefore, 35 VLEDs are required. Each VLED is a semiconductor PN junction diode interconnected as shown in *Figure 5-7*. It is quite easy to see that a particular diode can be turned on by selecting an X line (ROW) and a Y line (COLUMN).

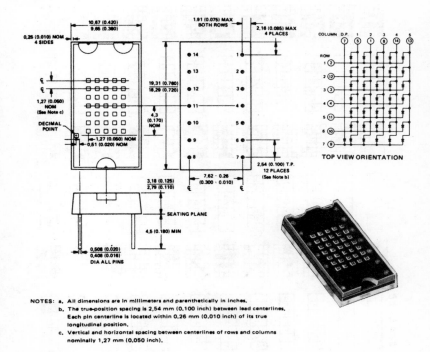

NOTES: a. All dimensions are in millimeters and parenthetically in inches.
b. The true-position spacing is 2,54 mm (0.100 inch) between lead centerlines. Each pin centerline is located within 0.26 mm (0.010 inch) of its true longitudinal position.
c. Vertical and horizontal spacing between centerlines of rows and columns nominally 1,27 mm (0.050 inch).

Figure 5-7. 5x7 Alphanumeric Display

OPTOELECTRONIC DISPLAYS 5

The amount of information which could be displayed by this design is tremendous. In fact, 3.44×10^{10} different configurations could be selected from such a display. However, since the purpose of the display is to communicate to people, only those configurations that represent characters which people are likely to recognize need to be displayed. The characters shown in *Figure 5-8* are representative of the 96 standard characters often used in computers. The ASCII code shown in *Figure 5-9* uses a 7-bit binary

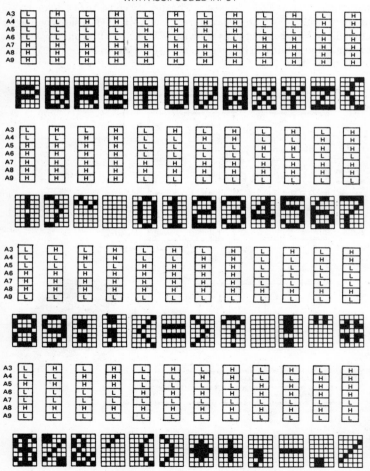

Figure 5-8. *Partial Character Set Generated from ASCII Coded Inputs*

5 OPTOELECTRONIC DISPLAYS

code to represent the 96 characters plus 32 control codes. When the ASCII coded input is applied to the seven input lines of the circuit shown in *Figure 5-10*, the TIL305 will display characters as shown in *Figure 5-8*.

All of the indicator, numeric and alphanumeric displays that have been shown are semiconductor PN junction light-emitting diodes made from compounds of gallium arsenide which may or may not include gallium phosphide. They can be obtained to have wavelength emissions that provide a yellow, red, or green color. They are packaged in many varieties of packages from single units to multiple unit arrays. However, other technologies are also used to provide these type displays. Let's look at several of these.

BIT POSITIONS:

7				0	0	0	0	1	1	1	1
6				0	0	1	1	0	0	1	1
5				0	1	0	1	0	1	0	1
4	3	2	1								
0	0	0	0	NUL	DLE	SP	0	@	P	\	p
0	0	0	1	SOH	DC1	!	1	A	Q	a	q
0	0	1	0	STX	DC2	"	2	B	R	b	r
0	0	1	1	ETX	DC3	#	3	C	S	c	s
0	1	0	0	EOT	DC4	$	4	D	T	d	t
0	1	0	1	ENQ	NAK	%	5	E	U	e	u
0	1	1	0	ACK	SYN	&	6	F	V	f	v
0	1	1	1	BEL	ETB	'	7	G	W	g	w
1	0	0	0	BS	CAN	(	8	H	X	h	x
1	0	0	1	HT	EM	)	9	I	Y	i	y
1	0	1	0	LF	SUB	*	:	J	Z	j	z
1	0	1	1	VT	ESC	+	;	K	[	k	{
1	1	0	0	FF	FS	,	<	L	\	l	|
1	1	0	1	CR	GS	-	=	M	]	m	}
1	1	1	0	SO	RS	.	>	N	∧	n	~
1	1	1	1	SI	US	/	?	O	—	o	DEL

EXAMPLES:

1000011 = C
0110011 = 3
1010000 = P
0110000 = 0 (ZERO)

SAMPLE OF CONTROL CHARACTERS (BOLD)

STX = Start of text
EOT = End of transmission
CR = Carriage return
SP = Space
HT = Horizontal tabulation

Figure 5-9. ASCII—American Standard Code for Information Interchange

OPTOELECTRONIC DISPLAYS 5

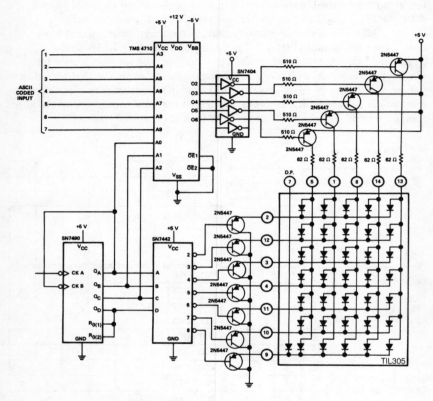

Figure 5-10. *Decode/Driver Circuit for ASCII Coded Inputs*

Liquid Crystal Displays

A display type that has unique features is the liquid crystal display. An LCD is used in the calculator shown in *Figure 5-11*. In some cases, considering this as an optoelectronic display may seem to be a misnomer because the liquid crystal display does not actually emit any light; it just uses the available light surrounding it to convey information. However, light is a major ingredient, the information displayed is controlled by electrical bias, and the content of the display is numeric and alphanumeric for visual communications. Therefore, the LCD is an optoelectronic display.

The basic principle of operation of a liquid crystal display is that it uses an electric field produced by applying a voltage to conducting plates on each side to control a medium through which light passes.

5 OPTOELECTRONIC DISPLAYS

Figure 5-11. Calculator with Liquid Crystal Display

The liquid crystal material (medium) is sandwiched between two glass plates which have a clear conducting material arranged in a pattern on their inner surfaces. When no bias is applied to the plates, the liquid crystal molecules represented by the oval shapes in *Figure 5-12a* are oriented such that light passes through the material with a minimum of scattering. The material appears clear. When a bias is applied, the crystals between the conductive plates are disoriented so that light is scattered as shown in *Figure 5-12b*. The result is a frosted appearance through the region of the display to which the bias is applied, (the region to the right in *Figure 5-12b*). If a black plate is placed behind the display, it absorbs the light that passes through so the area without bias appears black to the viewer. The result is a visible contrast caused by the different reflectivity at the right and left sides of the display in *Figure 5-12b*. By proper patterning of the biasing plates into segments or matrices, characters can be formed by using control circuitry similar to that described for VLED displays.

OPTOELECTRONIC DISPLAYS

5

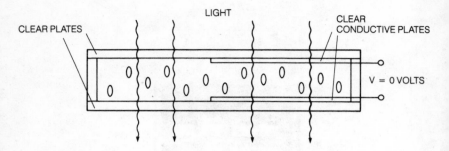

a. The Oval Shaped Crystals Are Aligned So That Most Of The Light Passes Through The Display.

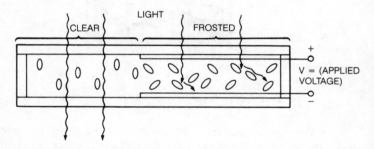

b. The Light On The Left Of The Display Passes Through While The Light On The Right Is Partially Blocked By The Randomly Distributed Crystals.

Figure 5-12. *Liquid Crystal Display*

The major advantages of these displays are that they consume very little power (a few microwatts) because they produce no light and they require very low voltages to operate. Thus, they can be driven directly from almost any low voltage logic circuit, multifunction integrated circuit, or microprocessor. The major disadvantages are that ambient light must be present so that the contrast can be recognized and performance is adversely affected by temperature. Cold temperature slows the response time and high temperature can cause a loss of the scattering produced by the bias. In many applications, the temperature performance is acceptable, but ambient light must be supplied to be able to see liquid crystal displays in the dark.

5 OPTOELECTRONIC DISPLAYS

Gas Discharge Displays

Another display type is the gas discharge unit. As shown in *Figure 5-13*, a reddish-yellow light is produced in the cathode region when the neon gas ionizes. The gas is usually neon with a small amount (about 3%) of argon added to reduce the voltage required to ionize the gas. A small amount of radioactive gas (krypton 85 isotope or tritium) may be added to provide ions which aid in starting the ionization process. Liquid mercury in small amounts also may be added to extend tube life by slowing the erosion of the cathode by the bombardment of relatively heavy positive ions.

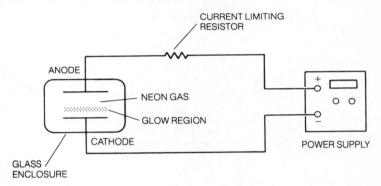

Figure 5-13. Gas Discharge Glow Tube

A typical one-digit display is shown in *Figure 5-14*. The basic package is an evacuated glass tube. The cathode bars are arranged to form the familiar seven-segment display configuration. The top of the display (anode) is either a metal screen (*Figure 5-14a*) or a transparent conductor (*Figure 5-14b*) spaced one or two millimeters away from the cathodes. Leads for the anode and each of the seven cathodes are brought outside the package and the unit is filled with a neon-argon gas mixture.

The voltage required to ionize the gas ranges from 100 to 200 volts. Since the display is usually driven by a digital logic circuit that has an output voltage much lower than this, some type of interface circuit to translate between voltage levels is usually required. A simple transistor drive circuit is shown in *Figure 5-15*. When V_1 is zero, transistor Q_1 is switched off. This allows the 180 volt supply to turn on the display through the 100 kilohm current limiting resistor. When V_1 is negative 5V, Q_1 is turned on and the voltage at the display cathode drops to 73 volts. This voltage is below the ionization potential required for the display so the display is turned off. The voltage required to turn the transistor on and off is a relatively low voltage; however, the transistor elements must be able to withstand the high ionization voltage without breakdown.

OPTOELECTRONIC DISPLAYS 5

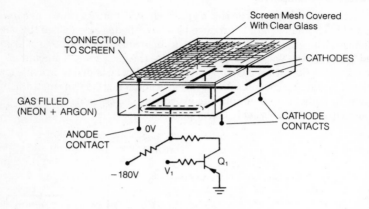

a. Construction Showing Screen Mesh Anode

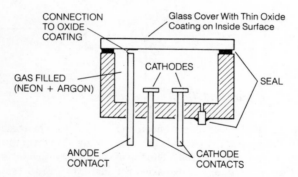

b. Cross Section Showing Transparent Oxide Coated Anode

Figure 5-14. *Typical Construction of Gas Discharge Displays*

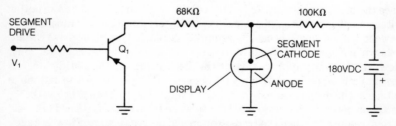

Figure 5-15. *Simple Transistor Driver for Gas Discharge Display*

5 OPTOELECTRONIC DISPLAYS

The principal advantage of the gas-discharge display is adequate brightness. The main disadvantage is the expense of the power supply and drive circuitry; however, another disadvantage for some applications is the possibility of the glass-enveloped display breaking under vibration or impact.

Vacuum Fluorescent Displays

The vacuum fluorescent (VF) display is basically a vacuum triode with phosphor-coated anodes. *Figure 5-16* illustrates a simple vacuum tube triode (three electrodes). The tungsten filament is coated with barium or thorium oxide which releases electrons when heated to approximately 700°C. The grid is a fine conductive mesh which is essentially transparent to the electron beam. The plate is a conductive surface which is the anode of the device. As the filament is heated, electrons are "boiled off" from the filament and form an "electron cloud" in the vicinity of the cathode. If a positive voltage is applied both to the plate and the grid, the electrons from the cathode will be drawn to both the grid and the anode so that both grid current and anode current flows as shown in *Figure 5-17a*. However, if the grid voltage is highly negative, the electrons will be repelled by the grid and, as a result, cannot get to the anode. Thus both the grid and anode currents are zero as shown in *Figure 5-17b*. Another possibility of interest is if the anode voltage is zero and the grid voltage is positive. In this case, a grid current flows, but the anode current is zero as shown in *Figure 5-17c*. The important thing to observe is that either the grid voltage or the anode voltage can be used to make the anode current zero.

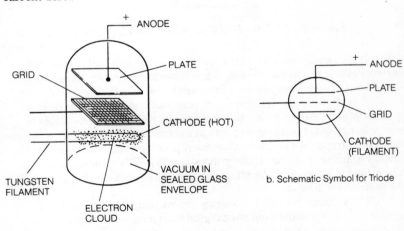

Figure 5-16. Vacuum Tube Triode

UNDERSTANDING OPTRONICS

OPTOELECTRONIC DISPLAYS

5

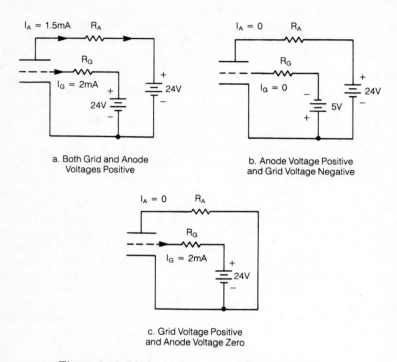

Figure 5-17. Methods of Controlling Anode Current

If the anode is coated with a phosphor, it will emit light when the electrons strike it or, in other words, when there is anode current the anode will emit light. These ideas are used to make the vacuum fluorescent display.

A typical vacuum fluorescent display unit is shown in *Figure 5-18*. Two filaments are used to make sure that enough electrons are available to reach each segment. Each segment acts as a separate anode while the grid controls the current to all of the segments. A positive voltage is applied to the segments which are to be on. If the grid voltage is positive, these segments will be on and the others will be off. If the grid voltage is zero or negative, all of the segments will be off.

The voltages required to operate the vacuum fluorescent display are larger than those available from most digital logic integrated circuits used. Although the voltages are much lower than the voltages required for the gas discharge displays, special interface circuits are still required. The color of the light from the display is usually a pleasing blue-green color. The intensity is good, but low contrast in ambient light can be a disadvantage.

5. OPTOELECTRONIC DISPLAYS

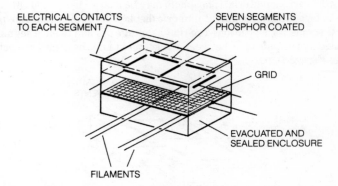

Figure 5-18. Simplified Construction of Vacuum Fluorescent Display

Electroluminescent Displays

Electroluminescent displays generate light when an electric field is applied to an electroluminescent phosphor. Most electroluminescent materials require activators. Activators are impurities in the material which determine the characteristics of the emitted radiation. A typical electroluminescent phosphor is zinc sulfide activated with manganese. *Figure 5-19* shows a cross section of an electroluminescent panel. The electroluminescent phosphor is sandwiched between two electrodes with insulating layers between the electrodes and the phosphor. One of the electrodes is a transparent conducting film. If an ac voltage is applied between the two electrodes, light is emitted through the transparent electrode.

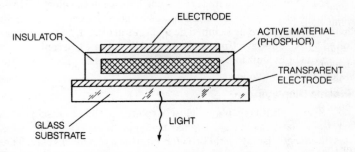

Figure 5-19. Cross Section of Electroluminescent Panel

OPTOELECTRONIC DISPLAYS

5

The active material (phosphor) may be shaped in several smaller regions to form a seven-segment numeric display. A set of seven transparent electrode segments is positioned so that a voltage applied to each segment activates only the phosphor between the segment and the common electrode as shown in *Figure 5-20*.

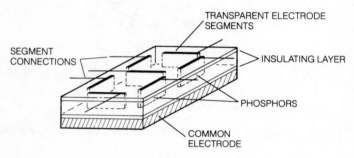

Figure 5-20. Typical Construction of Seven-Segment Electroluminescent Display

Electroluminescent panels may be made by evaporating aluminum onto glass. Patterns are then screened onto the aluminum surface. Insulating layers and phosphors are applied and the other electrode added. The electroluminescence material may be a powder or a thin film (4,000 Å thick). Clear plastic may be used instead of glass to produce a display that is light weight and also flexible to form either flat or curved viewing surfaces.

The display drive voltage may be ac or dc and is a moderately high voltage. Thinner films allow lower drive voltages to be used.

Since an insulating material is used between the electrodes and the phosphor, the electroluminescent displays behave electrically much like a capacitor. When pulsed or scanned dc voltages are used, the scanning rate is limited by the effective time constants in the circuit. Currents tend to be higher and a slight shift in color may be observed as the pulse repetition rate is increased.

AC Plasma Displays

The ac plasma display is closely related to the gas discharge display. In fact, the ionized gas in a gas discharge display is also called a plasma, so that the gas discharge display could be referred to as a plasma display. The ac plasma display uses the same type construction as the gas discharge display with an insulator added to both electrodes as shown in *Figure 5-21*.

5 OPTOELECTRONIC DISPLAYS

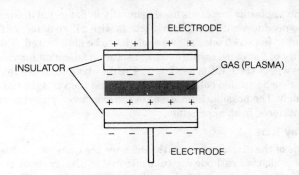

Figure 5-21. AC Plasma Display Cross Section

A relatively large ac drive voltage is applied between the two electrodes but the drive voltage is below the ionization potential of the gas so there is no light output as shown in *Figure 5-22*. A starting pulse with a voltage above the ionization potential is applied to ionize the plasma. When the gas ionizes, the insulators charge like small capacitors so that the sum of the drive voltage and the capacitive voltage is large enough to sustain the plasma. The current flow is determined by the width of the starting pulse; that is, the brightness can be controlled by the starting pulse. The display is turned off by momentarily interrupting the drive voltage.

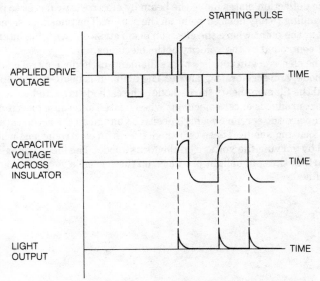

Figure 5-22. AC Plasma Display Driving Waveforms

OPTOELECTRONIC DISPLAYS

The ac plasma display is used primarily in flat-panel dot-matrix displays and produces satisfactory brightness with a high density of dots. The arrangement of the electrodes allows each dot to be illuminated. This leads to the possibility of flat panel displays with resolution adequate for video applications where the CRT is used now. The disadvantages are that a driver is needed for each row and column of the array and the voltages required are relatively high. The possible limitations of the evacuated glass package again must be considered in an application.

Cathode Ray Tube

One of the display types with which we are quite familiar is the TV picture tube which is a cathode ray tube adapted to display pictures. Advantages of this display are its large display area and versatility. The major disadvantage is that it requires complex and expensive support electronics circuitry, portions of which must operate at very high voltages. A secondary disadvantage might be the large-volume highly-evacuated glass tube required. Special precautions must be taken in case it breaks.

Figure 5-23 shows a basic cathode ray tube used in an oscilloscope, a laboratory instrument used to display waveforms for measurement in an electronic circuit. These waveforms are complex and fast changing voltages that cannot be accurately measured by conventional voltmeters.

The CRT shown in *Figure 5-23* consists of three main functions: the electron gun, the deflection plates, and the phosphor screen. The electron gun produces a high velocity, focused beam of electrons that are "fired" toward the screen. The deflection plates move the beam by electrostatic forces to position the beam landing to any desired point on the screen. The phosphor screen emits light at the point where the electron beam strikes it with the intensity of the light controlled by the velocity of the electrons.

The electron gun consists of the filament, cathode, control grid, focusing anode, and accelerating anode. The filament is heated by an applied voltage and the filament heats the cathode. The cathode is coated with a material that produces excess electrons when heated and these electrons are attracted toward the screen by a high positive voltage on the accelerating anode and focusing anode. These two anodes form an electronic lens which is adjusted by varying the voltage on the focus anode. The focus voltage is adjusted so that the spot of light produced on the screen is very small and sharply defined.

5. Optoelectronic Displays

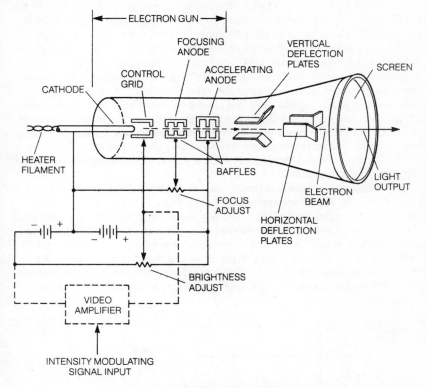

Figure 5-23. Oscilloscope Cathode Ray Tube

The control grid action of the CRT is very much like that explained for the vacuum fluorescent display tube in *Figure 5-17*. The control grid regulates the velocity and number of electrons that are fired from the gun. This is accomplished by varying the voltage on the control grid much like you vary the stream of water from a faucet by turning the valve handle. Therefore, the electron beam can be intensity modulated by varying the control grid voltage. More about this later.

The electron beam from the gun must pass between the two sets of deflection plates on the way to the screen. Varying the voltage difference between the vertical deflection plates moves the beam, and the spot of light on the screen, up and down. Likewise, varying the voltage difference between the horizontal deflection plates moves the spot of light from side-to-side on the screen. If the voltage on both sets of deflection plates is varied at the same time, the spot of light can be moved to any position on the screen.

OPTOELECTRONIC DISPLAYS 5

The screen is coated with an electro-luminescent material such as zinc sulfide which emits light when excited by the striking electrons. The color of the emitted light depends on the particular material used. An oscilloscope CRT usually emits green or blue light. Some video terminal CRTs emit green light while others emit white light. Television picture tubes that can only produce black and white pictures are CRTs that emit white light. TV picture tubes that produce pictures in color have three different phosphor types arranged in a particular pattern. The phosphors emit the primary light colors of red, blue, and green and other colors are produced by the mixing of correct proportions of these three colors. All phosphors used as a CRT screen coating also have the characteristic of emitting light for a short time after the electron excitation is removed. This characteristic is useful when the screen is scanned as will be explained below.

The oscilloscope is normally used to display voltage variations with respect to time. This is accomplished by applying a voltage that varies at a constant rate, called the sweep, to the horizontal deflection plates so that the spot moves from one side of the screen to the other at a constant speed. The waveform or voltage variation to be observed is applied through appropriate amplifiers to the vertical deflection plates as illustrated in *Figure 5-24*. As the spot is swept across the screen, the waveform voltage moves the spot up and down so that the spot of light traces an exact pattern of the waveform on the screen.

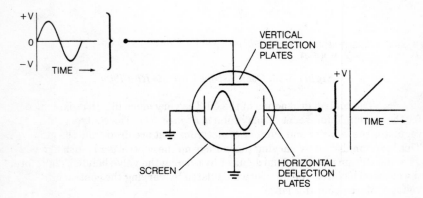

Figure 5-24. Oscilloscope Waveform on CRT Screen

5 OPTOELECTRONIC DISPLAYS

The CRT used as a television picture tube does not have deflection plates because the spot deflection is accomplished by varying the current through a set of vertical deflection coils and a set of horizontal deflection coils. These coils are placed externally around the neck of the CRT to produce magnetic deflection. Another difference for CRTs that produce pictures in color is that most of these CRTs have three electron guns, one for each of the three color phosphors on the screen.

The CRT in a TV receiver or video terminal is also operated differently from the oscilloscope. In a TV, both vertical and horizontal deflection is continuously swept by voltages (or currents) at a constant rate. However, the two rates are different, with the horizontal sweep speed about 263 times faster than the vertical sweep speed. The result is that the spot is swept or scanned very rapidly across the screen and slightly downward so that the entire screen is "painted" one line at a time. The sweep rate is so fast that the screen is completely scanned 30 times in one second. This, combined with the phosphor characteristic of emitting light for a short time after excitation is removed, and the human eye's characteristic of continuing to "see" light for a short time after the light goes out, gives the illusion that the screen is continuously lighted. The scanned and lighted screen with no information present is called a raster.

Another difference in operation between the oscilloscope and the TV is the use of the control grid for intensity modulation. Some oscilloscopes provide an input for this purpose but it is not used very much. Instead, an intensity or brightness control varies the voltage on the control grid to adjust the brightness of the trace (the moving spot of light) as desired and then the intensity stays constant. A brightness control is also used on a TV to adjust the brightness of the raster, but the video signal which contains the information to reproduce the transmitted picture is also applied through the video amplifier to the control grid (*Figure 5-23*). As the video signal voltage varies up and down, it changes the flow of electrons from the gun. The varying electron flow varies the brightness of the spot on the screen as the spot is being swept by the deflection coils. Synchronization signals keep the video signals and sweep signals in time so that the beam intensity varies at the correct location on the screen. The result is that the transmitted picture is reproduced on the screen in continuously varying tones of gray ranging from black to white for a black and white picture. Color pictures are reproduced similarly but the intensity of each of the three guns is controlled separately.

CRT displays used for a computer terminal (*Figure 5-25*) and other types of information displays also use the control grid for intensity modulation, but usually the beam is turned on or off fully because the video signal is a digital signal that jumps between the low and high logic levels. The result is a high contrast ratio display of numeric, alphanumeric, or graphic information from the keyboard input or from the computer output of calculations and data.

OPTOELECTRONIC DISPLAYS 5

Figure 5-25. Typical Computer Terminal

Ink Jet Printer

An interesting extension of CRT technology is the ink jet printer, shown schematically in *Figure 5-26*. Essentially the heater filament and cathode of the CRT are replaced by an ink gun. The ink is electrically charged and the stream of ink is controlled in much the same way as the electron beam of the cathode ray tube. A sheet of paper replaces the screen and characters are "drawn" on the paper. The result is high-speed and noiseless printing that provides a paper and ink display of good quality.

WHAT ARE THE IMPORTANT CHARACTERISTICS OF DISPLAYS

The important characteristics of displays are viewability, response time, power requirements, and reliability.

Viewability

Viewability requires that the light emission be in the visible range, that it be bright enough to be distinguishable from its surroundings in the expected ambient lighting, and that it be legible from the viewing angle that its application requires.

Although the viewing angle is an important consideration, it is not always specified in data sheets. Because the viewability of any display may vary widely depending upon the application, a judgment should be made as to its suitability by actually viewing the display at the extremes of the angles expected in the application.

5 OPTOELECTRONIC DISPLAYS

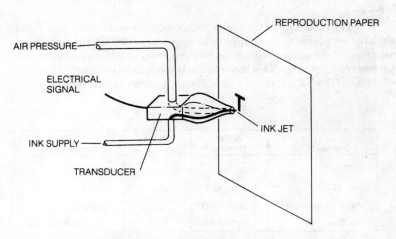

Figure 5-26. Basic Ink-Jet Printing Technique

Response Time

The response time of a display becomes important when it is coupled to a fast changing signal such as digital data. However, a simple fault or status indicator is usually always either on or off and just switches states once for its indication of status. Therefore, for indicator applications, switching speed is not very important.

Power Requirements

The power requirements of displays vary widely and typical values are presented in *Table 5-2*. One can see why the LCD display is used so extensively in hand-held calculators and digital watches. It has a 4 to 1 advantage in power over its nearest competitor.

Table 5-2. Power Requirements

Display Type	Volts	Milliamperes
VLED	2	10
Liquid Crystal	5	1
Vacuum Fluorescent		
Anode	35	10
Filament	1.5	10
Gas Discharge	90	2

OPTOELECTRONIC DISPLAYS

Reliability

The reliability of the major technologies is quite adequate. *Table 5-3* tabulates approximate mean time between failures (MTBF) and shock resistance. Solid-state construction has improved reliability of electronic systems tremendously. This is particularly evident by comparing the reliability of the VLED to the reliability of the other technologies.

Table 5-3. Reliability

Display Type	MTBF (Hours)	Shock Resistance	Construction
VLED	>20000	High	Solid State
Liquid Crystal	5000-15000	Fair	Glass
Vacuum Fluorescent	5000-15000	Fair	Glass Sealed Panel
Gas Discharge	5000-15000	Fair	Glass Sealed Panel

Table 5-4 compares features of the display types that have been mentioned. Notice that all of the display types have acceptable visibility in dim ambient light; however, in direct sunlight only the LCD and CRT are considered good. Viewability of the other types may be improved with the use of optical filters to improve contrast ratio. With the exception of the LCD, all display types have good response time, but very fast response time is available only with the VLED and the CRT. Power requirements are moderate for most display types with the LCD being a good choice for battery operated units because of its very low power consumption.

Table 5-4. Comparison of Display Types

Display Type	Viewability		Response Time	Power Requirements	Reliability	Voltage Requirements
	In Dim Light	In Direct Sunlight				
VLED	Excellent	Poor	Excellent	Moderate	Excellent	Low
LCD	Good	Good	Poor	Very Low	Good	Low
Gas Discharge	Excellent	Poor	Good	Moderate	Good	Moderate
AC Plasma	Excellent	Poor	Good	Moderate	Good	Moderate
Vacuum Fluorescent	Excellent	Poor	Good	Moderate	Good	Moderate
Electroluminescent	Excellent	Poor	Good	Moderate	Good	Moderate
CRT	Excellent	Good	Excellent	High	Fair	High

5 OPTOELECTRONIC DISPLAYS

WHAT ARE SOME APPLICATIONS OF DISPLAYS?
Calculators

Calculator displays fall into two categories; those used in portable calculators and those used in desk-top calculators. Because of the need to conserve power in portable calculators, only two display types have been commonly used: the VLED and liquid crystal. Liquid crystal has begun to dominate in recent years and is used exclusively in the solar powered calculators.

Vacuum fluorescent and gas discharge displays have had limited use in portable applications because of their greater power requirements. On the other hand, these displays dominate the desk-top calculators because of their excellent viewability due to the bright and large display characters. Power conservation is not a major requirement since they are normally powered by plugging into the ac line outlet.

It should also be observed that hand-held calculators usually need only numeric displays with a few symbols. Hand-held computers, however, may need alphanumeric displays such as the dot-matrix arrays.

Watches

Only two display types have found use in watches. Because of the small space allowed for a battery, it is important that the display not consume much power. The LED display requires so much power that the battery life would be very short if the display were always on. Therefore, the user is required to press a switch to turn on the display only when needed. However, the power requirement of the liquid crystal display is so low that it can display the time constantly and still have long battery life. Since it is legible even in bright sunlight, it has become dominant for the digital watch display.

Because the liquid crystal display is not visible in the dark, most LCD watches have a built-in illumination source which is often a tiny incandescent bulb. Since it requires a relatively large amount of power, it is normally off and is switched on by the user when necessary. Tritium, a source of light radiation which requires no power, is used for backlighting in some LCD watches. It deteriorates with time but remains effective for years.

Watch displays usually need only numeric displays with a few special symbols for alarms and days of the week.

OPTOELECTRONIC DISPLAYS 5

Electronic Games

Because of the wide variety of applications in electronic games requiring fast motion and unusual display formats, LEDs have remained popular in physically small electronic games. LCDs; however, still have an advantage because of their low power demand and may eventually dominate this field. The CRT (picture tube) of the television receiver is ideal for stationary applications because it is readily available, has a large viewing area, can be in full color, and has fast response time.

WHAT HAVE WE LEARNED?

1) Displays of many different technologies have been developed to make visible the increasingly complex information available to people.
2) Displays, which range in complexity from simple indicators to detailed pictures, bring information about something to people.
3) Displays have been specialized according to a required function: on-off indicators, numeric, alphanumeric, and graphical (pictures).
4) The most important characteristic of a display is that it present the message to the eye so that it may be quickly and correctly interpreted.
5) Solid-state technology has increased the reliability of displays a great deal.
6) VLEDs are used for highly reliable solid-state displays that have fast response time and reasonably low power.
7) Liquid crystal displays are used in battery operated equipment because of the low power consumption.
8) CRT displays are used because of their availability, large viewing area, fast response time, and color capability.

5 OPTOELECTRONIC DISPLAYS

Quiz for Chapter 5

1. What is the purpose of a display?
   a. To look pretty
   b. To illuminate something
   c. To make visible what cannot be seen
   d. To bring information to people

2. What is the main advantage of the CRT?
   a. Versatility
   b. Power requirements
   c. Brightness
   d. Portability

3. What are the dominant requirements of a watch display?
   a. Brightness
   b. Contrast ratio
   c. Viewing angle
   d. Power requirements

4. How many different things can a 4-element linear array indicate?
   a. 4
   b. 25
   c. 20
   d. 16

5. Contrast ratio is the ratio of
   a. display brightness to background brightness.
   b. display color to background color.
   c. height to width.
   d. none of the above.

6. The lowest power requirement display discussed in this chapter is the
   a. CRT.
   b. LCD.
   c. VF.
   d. LED.

7. A feature common to all displays is the
   a. emission of light.
   b. control of light.
   c. display of information.
   d. none of the above.

8. The display type which requires the lowest voltage is the
   a. liquid crystal.
   b. cathode ray tube (CRT).
   c. VLED.
   d. vacuum fluorescent.

9. The display type with the highest reliability and best shock resistance is the
   a. liquid crystal.
   b. CRT.
   c. VLED.
   d. vacuum fluorescent.

10. Displays are multiplexed in order to
    a. reduce power requirements.
    b. require fewer data lines.
    c. reduce the number of connections to the display.
    d. all of the above.

11. In a seven-segment display, the number 4 is represented when segments _____ are on.
    a. a, b, c, d, e, f
    b. b, c, f, g
    c. c, d, e, f, g
    d. b, e, f, g

12. The number of possible displays available within a seven-segment display is
    a. 7
    b. 10
    c. 64
    d. 128

13. A 5 x 7 dot matrix array is used for
    a. numeric displays only.
    b. alphanumeric display only.
    c. words only.
    d. symbols only.

14. The ASCII code is a seven-bit code used to generate
    a. characters only.
    b. control codes only.
    c. characters and codes.
    d. numbers only.

OPTOELECTRONIC DISPLAYS 5

15. The number of electrons in the CRT beam is controlled by
    a. varying the voltage on the horizontal deflection plates.
    b. varying the voltage on the control grid.
    c. varying the voltage on the vertical deflection plates.
    d. the phosphor screen.

16. The electron beam of a CRT must scan the screen in order to
    a. refresh the display.
    b. reduce flicker.
    c. illuminate the background of the whole screen.
    d. all of the above.

17. For the SN7447A decoder/driver of *Figure 5-2*, if the inputs are LT = RB1 = BI/RBO = H (high), and A = B = C = L (low) and D = H (high) the output will light segments to display.
    a. 0
    b. 1
    c. 4
    d. 8

18. The ripple blanking input of the TIL306 (*Figure 5-5*) allows for digit display of the number 12 to be displayed as
    a. 0012
    b. 12
    c. 1200
    d. 012

19. Gas discharge displays may contain a small amount of krypton 85 isotope or tritium to
    a. slow cathode erosion.
    b. extend tube life.
    c. aid in starting ionization process.
    d. control the color.

20. The CRT (picture tube) in a television set differs from the CRT in an oscilloscope in that it
    a. has magnetic deflection coils.
    b. the beam continuously scans the screen.
    c. the control grid modulates the beam.
    d. all of the above.

1. d, 2. a, 3. d, 4. d, 5. a, 6. b, 7. c, 8. c, 9. c, 10. d, 11. b, 12. d, 13. b, 14. c, 15. b, 16. d, 17. d, 18. b, 19. c, 20. d

6 APPLICATIONS OF LIGHT-EMITTING DIODES

Applications of Light-Emitting Diodes

ABOUT THIS CHAPTER

Because of the advantages of light-emitting diodes, LEDs have replaced miniature incandescent lamps in visible light applications which require high reliability, low cost, low power consumption, vibration resistance, fast response, and/or small size. LEDs are easy to use with both digital and analog semiconductor devices.

To make it easier to see how the characteristics are evaluated for an application, we've divided the characteristics into three sections: electro-optical, optical, and electrical.

WHAT ARE THE ELECTRO-OPTICAL CHARACTERISTICS OF A VLED?

The electro-optical characteristics provide information about the light output as a function of the current through the device. We can use data from a manufacturer's data sheet to examine these characteristics. *Figure 6-1* presents typical data for two gallium arsenide phosphide VLEDs. The two VLEDs are very similar except that one (TIL220) has a red molded body while the other (TIL221) has a colorless body. *Figure 6-1a* gives the operating characteristics at 25° C. Notice that the typical wavelength is 650 nanometers or 6,500Å, which is a deep visible red.

Figure 6-1b shows the relationship between the relative luminous intensity and the forward diode current. The luminous intensity is given in relative terms so that the same graph can be used for both devices. The scale can be interpreted in percent with 1 being 100 percent. To use this graph, select a forward current and draw a vertical line upward until it intersects the curve; then draw a horizontal line from that point to the left and read the relative luminous intensity from the vertical axis. For example, if I_F is 20mA, the relative luminous intensity is 1 or 100% of the 800 microcandelas given in *Figure 6-1a* for the TIL220 or 100% of the 1000 microcandelas specified for the TIL221. If the current is reduced to 10mA the relative luminous intensity drops to 0.5 or 50% of the value given in *Figure 6-1a*.

The luminous intensity in *Figure 6-1b* is specified at a room ambient temperature (T_A) of 25°C (77°F). *Figure 6-1c* shows that luminous intensity decreases as temperature increases.

The wavelength will also shift with temperature. In most applications of visible LEDs the shift is not noticeable to the eye. However, the shift of wavelength could cause problems in some applications of infrared LEDs where the detector's spectral response must be considered. This shift (an increase) in wavelength is in the order of 2.5Å per degree centigrade increase in temperature.

APPLICATIONS OF LIGHT-EMITTING DIODES

6

operating characteristics at 25° free-air temperature

PARAMETER		TEST CONDITIONS	TIL220			TIL221			UNIT
			MIN	TYP	MAX	MIN	TYP	MAX	
I_V	Luminous Intensity	$I_F = 20$ mA	800			1000			μcd
λp	Wavelength at Peak Emission	$I_F = 20$ mA	6300	6500	6700	6300	6500	6700	Å
V_F	Static Forward Voltage	$I_F = 20$ mA		1.6	2		1.6	2	V
I_R	Static Reverse Current	$V_R = 3$ V		0.1			0.1		μA

a.

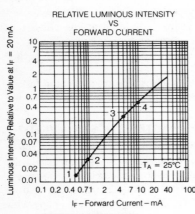

IN ROOM AMBIENT LIGHT
1 visible
2 easily seen
3 attract attention of casual observer
4 seen from 20 feet or more

b.

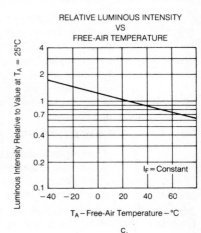

c.

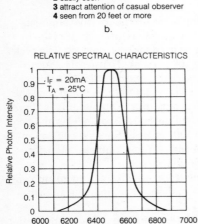

d.

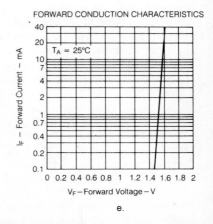

e.

Figure 6-1. *Typical Data for TIL220 and TIL221 Gallium Arsenide Phosphide Visible Light-Emitting Diodes*

6 APPLICATIONS OF LIGHT-EMITTING DIODES

WHAT ARE THE OPTICAL CHARACTERISTICS OF VLEDs?

The optical characteristics of the VLED include color (wavelength), viewing angle, and type of lens used with the device. *Figure 6-1d* shows the curve of light intensity as a function of wavelength. Notice that the spectrum is narrow with the maximum intensity at 6,500Å and only 10% of maximum intensity at 6,300Å and at 6,700Å.

Figure 6-2 presents typical data for two gallium phosphide VLEDs. Notice in *Figure 6-2a* that the wavelength is 6,200 angstroms. This color is a lighter shade of red than the TIL220.

operating characteristics at 25°C free-air temperature

PARAMETER		TEST CONDITIONS		MIN	TYP	MAX	UNIT
I_v	Luminous Intensity	I_F = 20 mA	TIL228-1	2.1			mcd
			TIL228-2	6			
			TIL231-1	6			
			TIL231-2	15			
λ_p	Wavelength at Peak Emission	I_F = 20 mA			6200		Å
θ_{HI}	Half-Intensity Beam Angle	I_F = 20 mA	TIL228		60°		
			TIL231		25°		
V_F	Static Forward Voltage	I_F = 20 mA				3.2	V
I_R	Static Reverse Current	V_R = 5 V				100	μA

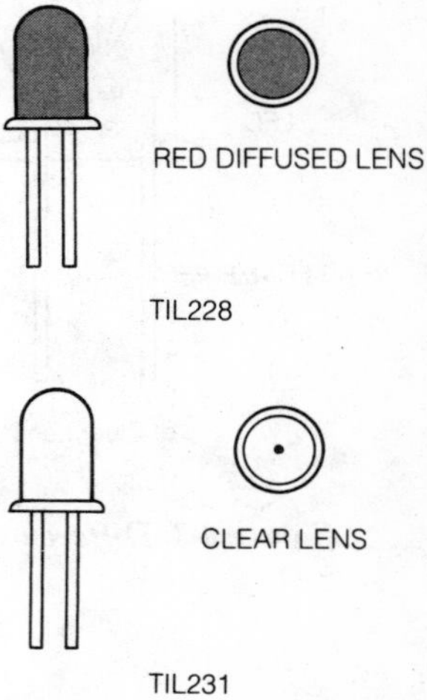

a.

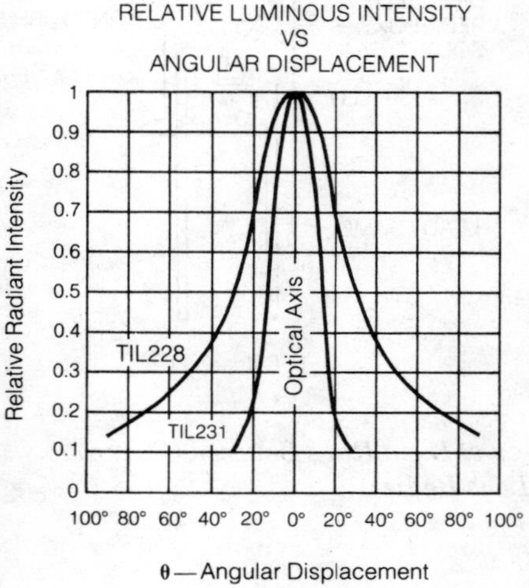

Figure 6-2. *Selected Characteristics of the TIL228 and TIL231 Gallium Phosphide Visible Light-Emitting Diodes*

APPLICATIONS OF LIGHT-EMITTING DIODES

Figure 6-2b shows the relative luminous intensity as a function of angle for the two devices. Although the same chip and same package dimensions are used, it is obvious that the TIL23 has a narrower viewing angle than the TIL228. Also notice from *Figure 6-2a* that the TIL231-1 produces a luminous intensity of 6 millicandelas (6,000 microcandelas) compared to 2.1 millicandelas (2,100 microcandelas) for the TIL228-1. *Figure 6-2c* illustrates the view of the two diodes from the optical axis. The TIL228 has a red filled epoxy lens and produces almost uniform light over the surface of the lens. The TIL231 has a clear epoxy lens of the same dimensions and provides a small spot of light of higher intensity than the TIL228.

The reason for the higher light intensity of the TIL231 is illustrated in *Figure 6-3*. In *Figure 6-3a*, the light emitted from the chip is transmitted through the clear plastic body of the lens and is focused into a narrow 25 degree viewing angle. In *Figure 6-3b*, the light emitted from the chip is reflected inside the red plastic epoxy body so that the light rays leave the lens in many directions and produce a wider viewing angle.

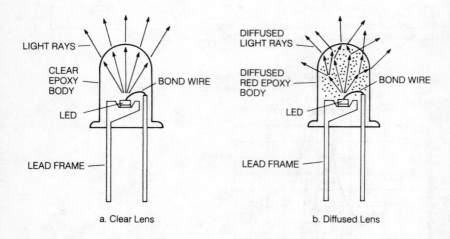

Figure 6-3. *Difference in Dispersion of Light Rays Between Clear and Diffused Lens Bodies*

6 APPLICATIONS OF LIGHT-EMITTING DIODES

WHAT ARE THE ELECTRICAL CHARACTERISTICS OF AN LED?

Electrical characteristics of the LEDs are similar to those of any semiconductor diode. Ratings such as reverse voltage, reverse current, forward voltage, and maximum continuous forward current are usually specified at 25°C free-air temperature. Typical data is shown in *Figure 6-1a* and *Figure 6-2a*. Notice on the forward conduction characteristics curve depicted in *Figure 6-1e* that the forward voltage varies only a small amount for a large variation in forward current, from 1.45 volts at 0.1mA to 1.6 volts at 20mA. (The forward voltage of the TIL228 at 20mA is about 2.2 volts).

Since the light output depends on current, it is usually desirable to drive the LED from a controlled or constant current source. *Figure 6-4* shows typical drive circuits. Both circuits in *Figure 6-4a* and *b* operate in essentially the same manner. The current through the LED is set by the ratio of R_1 to R_2 and the size of the feedback resistor R_3. If the current through the LED tends to increase, the higher voltage drop across R_3 reduces the forward bias current through the base of the transistor and therefore stabilizes the current through the LED. The circuit of *Figure 6-4c* uses an operational amplifier in a constant current mode. The current through the LED depends only on the voltage (+V) and R_1. The current is independent of any changes in the LED due to temperature.

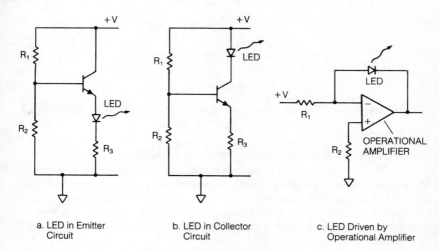

a. LED in Emitter Circuit

b. LED in Collector Circuit

c. LED Driven by Operational Amplifier

Figure 6-4. *Constant Current Drive Circuits for LEDs*

APPLICATIONS OF LIGHT-EMITTING DIODES 6

Power dissipation is computed by multiplying the forward voltage by the forward current. For example, if the TIL220 is operated at 20mA, the forward voltage is 1.6 volts and the power dissipation is 32 milliwatts.

Power dissipation is important because it causes the temperature of the device to increase. Remember that light intensity decreases and wavelength increases as temperature increases. Of course lower power saves energy but light output is the primary concern.

Other related data includes information about reliability. *Figure 6-5* shows typical reliability data for a gallium arsenide infrared LED. This graph shows that power output decreases as the operating time increases, and that the power output decreases faster if the operating current is higher. When the LED is operated at 50mA, it may lose as much as 20% of its original output power in 100,000 hours (11.4 years) of continuous operation; however, at 10mA the change in output power is insignificant in 100,000 hours of operation.

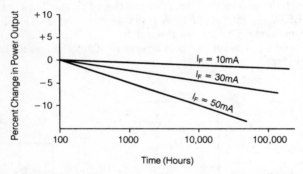

Figure 6-5. *Typical Data for Change in Output Power as a Function of Time for Light-Emitting Diodes*

Table 6-1 shows another important measure of reliability. If the LED suddenly stops emitting light, it is classified as a catastrophic failure. If the light level drops below performance specification but still provides light, it is referred to as a degradation failure. The failure rate is statistically calculated from test data. A failure rate of 0.28% per 1000 hours with a 90% confidence level implies a 90% certainty that no more than 0.0028 units will fail in 1000 hours or 2.8 units in 1 million hours.

6 APPLICATIONS OF LIGHT-EMITTING DIODES

Table 6-1. *Typical Failure Data for Light-Emitting Diodes*

Units Tested	Unit Hours	Catastrophic Failure	Degradation Failure Rate Failure Rate in %/1,000 hours			Temp °C	Current mA
			Total	60% Confidence	90% Confidence		
1384	1,384,000	0	1	0.15	0.28	25	50
432	432,000	0	2	0.72	1.23	25	75
432	432,000	0	3	0.97	1.55	55	50

The data from *Table 6-1* shows that in the test units there were no catastrophic failures and only six degradation failures. However, three of the six failures were in the last group which was tested at 55°C and 50mA. A comparison of the data shows that if the temperature is the same, the devices operated at higher current have higher failure rates. The data also shows that if the current is the same, the devices operated at higher temperatures have higher failure rates. Also, failure rates will be higher for high temperatures and moderate currents than failure rates for room temperature operation at higher currents. These conclusions are based on statistical data at normal operating temperatures and currents. It would be a good idea to obtain more data from the manufacturer if the devices are to be used at extremely cold temperatures and low current levels.

HOW ARE VLEDs SELECTED?

The selection of a VLED must begin with a definition of its intended use. Then, data sheets must be studied to determine the most likely candidates.

In order to illustrate this process, consider the following example. A panel containing the word "OIL" is to be illuminated when the oil pressure of an automobile is dangerously low. The output of the sensor circuit which makes the decision is a TTL device capable of sinking 10 milliamps. The TTL device will turn on the VLED. The electrical drive circuit is important, but let's look at the optical considerations first and later we'll see how the circuitry drives the VLED. The size of the word "OIL" is shown in *Figure 6-6a* and must be clearly visible when illuminated. For this example, the ambient light level is considered to be relatively low. The VLED is to be placed about an inch behind the panel to illuminate the panel from the back as shown in *Figure 6-6b*. Since the eye cannot discern a change in intensity of less than a 2 to 1 ratio, the illumination of the panel is allowed to vary up to a ratio of 2 to 1.

APPLICATIONS OF LIGHT-EMITTING DIODES

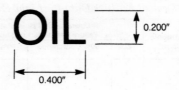

a. Size of Word

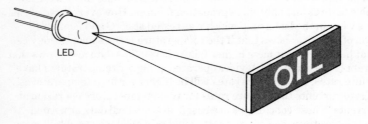

b. Method of Illuminating Word

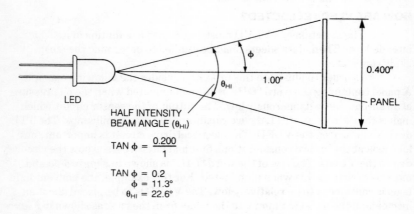

c. Calculating Half-Intensity Beam Angle

Figure 6-6. *Illuminating a Display Panel From the Rear*

6 APPLICATIONS OF LIGHT-EMITTING DIODES

The application requirement for the half-intensity beam angle (viewing angle) is shown and calculated in *Figure 6-6c*. The result shows that the half-intensity beam angle must be at least 22.6 degrees. Since the application is a warning to be detected by the eye, the traditional warning color (wavelength) of red is selected. Both the TIL228 and the TIL231 (shown in *Figure 6-2)* meet the criteria so far, since they both produce red light and have a half-intensity beam angle greater than 22.6 degrees. However, it is also desirable to produce the maximum amount of illumination. Referring again to *Figure 6-2*, the brightness of the TIL231-2 is 15 millicandelas (mcd) compared to only 6 mcd for the TIL228-2 at 20mA. At the anticipated operating current of 10mA the brightness would drop to 7.5 mcd and 3mcd respectively. Therefore, the TIL 231 is selected because its half-intensity beam angle of 25 degrees is adequate, its intensity is good, and it can be powered directly by the TTL device of the sensor circuit output.

Any application can be evaluated in much the same way, that is, the optics are matched to the application, the wavelength is matched to the sensor, and the brightness and electrical power are matched to the requirement.

The assumption was made in the preceding example that the ambient light level were relatively low. But what happens when it is not low? If unwanted light were shining from the VLED side of the panel, the warning "OIL" would appear to be on when it is not. The easiest way to avoid this is to enclose the LED side so that no light can enter except that from the LED. But what about the other side of the panel? In an automobile, the ambient light in the daytime is high on the viewer's side of the panel. The goal is to produce a display so that the word "OIL" is not legible *(as shown in Figure 6-7a)* when the VLED is off, and which is legible (as shown in *Figure 6-7c*) when the VLED is on. This can be accomplished by minimizing the reflected light from the display panel and optimizing the amount of light from the VLED which reaches the viewer through the illuminated word on the panel. One way to do this is to make the panel black and the word "OIL" a dark red so that ambient light is not reflected to the viewer's eye. The black surface absorbs all of the light and the red surface of the letters can be designed so it absorbs all but the narrow band of red emitted by the VLED. This minimizes the reflected light interference while optimizing the contrast ratio of the display when the VLED is on.

Figure 6-7 illustrates the effects of contrast ratio. Notice that *Figure 6-7c* is more legible than *Figure 6-7b* which is more legible than *Figure 6-7a*. From this, it is apparent that the design of the panel is also important to the application of the VLED.

APPLICATIONS OF LIGHT-EMITTING DIODES   6

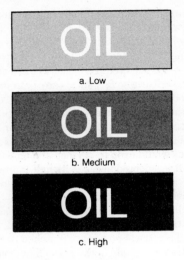

Figure 6-7. Contrast Ratio

Contrast ratio has been defined in many ways. A complete discussion can be very complex; however, contrast ratio as defined below is a useful concept in evaluating display performance.

$$\text{CONTRAST RATIO} = \frac{\text{Light from "OIL"} + \text{Light from background}}{\text{Light from background}}$$

This term may be applied when the VLED is on and when it is off. If there is no light from the word "OIL", the contrast ratio would be one. If the light from the word "OIL" is equal to the background light, the contrast ratio would be 2. The desire is that the contrast ratio be low when the VLED is off and high when the VLED is on. A measure of the readability of the display is the contrast improvement ratio defined below:

$$\text{CONTRAST IMPROVEMENT RATIO} = \frac{\text{Contrast Ratio (VLED on)}}{\text{Contrast Ratio (VLED off)}}$$

The contrast ratio can be measured using a light meter with a filter that approximates the response of the human eye. Two measurements should be taken. The first measurement is made of the display, the second is made of the display with the word "OIL" blacked out so that it reflects no light. The ratio of these two readings is the contrast ratio. The contrast improvement ratio is measured by reading the light meter when the display is "on" and dividing by the reading obtained when the display is "off".

6. APPLICATIONS OF LIGHT-EMITTING DIODES

As a general rule, contrast ratios less than 2 are difficult for the eye to detect while contrast ratios of 10 are easy to detect. This is also true for contrast improvement ratios. It is, therefore, desirable to design the panel so that the contrast ratio caused by ambient light when the VLED is off is less than 2 and the contrast ratio when the VLED is on is greater than 10. This would assure that the contrast improvement ratio is greater than 5. Since the contrast improvement ratio decreases as the ambient light increases, the contrast improvement ratio of 5 allows a design margin above the minimum of 2. Referring to *Figure 6-7*, (a) has a low contrast ratio, (b) has a medium contrast ratio, and (c) has a high contrast ratio. As you can see, the contrast improvement ratio between (a) and (c) is good.

WHAT ARE APPLICATIONS OF VLEDs?

Status Indicators

Status indicators serve a simple but very important function. As technology becomes more complex and the equipment that we work with becomes more sophisticated, it is more difficult to determine if the equipment is functioning properly. For example, in early aircraft, a pilot or copilot would turn a crank which was geared to the landing gear to lower the wheels. When the gear reached a positive stop, he knew that the landing gear was down. In modern aircraft, a switch is activated which in turn may activate a hydraulic system to lower the wheels. Since the pilot cannot see the landing gear and has no "feel" for it, he needs a status indicator to assure him that the landing gear is down. In other situations, modern systems such as computers are so complex that, even though we programmed them, it is difficult to feel confident that the system actually performed the task without some status information from the computer.

VLEDs are especially good for use in applications that require continuous monitoring because they are small and have low power requirements. *Table 6-2* is a typical quick reference guide which compares color, lens, brightness, and package size for VLEDs. Reference guides similar to this are available from most manufacturers. Since red and green colors traditionally mean stop and go or dangerous and safe conditions respectively, the use of these color VLEDs as status indicators is common. Status indicators can be used to indicate the presence or absence of a signal for information feedback to the operator. Status indicators also can be used to indicate an equipment malfunction or failure so the operator knows that corrective action must be taken. One example of a status indicator is the use for logic level indication in computers.

APPLICATIONS OF LIGHT-EMITTING DIODES 6

Table 6-2. *Quick Reference Guide for VLEDs*

Device	Color	Lens	Brightness MIN (mcd)@I_F (mA)		Package (Lamp Size)	Features
5082-4550	Yellow	Diffused	1	10		
5082-4555	Yellow	Diffused	2.2	10		
5082-4650	Red	Diffused	1	10	CL-10	
5082-4655	Red	Diffused	3	10	T1-¾	
5082-4950	Green	Diffused	1	20		
5082-4955	Green	Diffused	2.2	20		
TIL209A	Red	Diffused	0.5	20		
TIL211	Green	Diffused	0.8	25		
TIL212-1	Yellow	Diffused	0.8	20	CL-9	
TIL212-2	Yellow	Diffused	2.1	20	(T1)	
TIL216-1	Red	Diffused	2.1	20		
TIL216-2	Red	Diffused	6	20		
TIL220	Red	Diffused	0.8	20		
TIL221	Red	Clear	1	20		
TIL224-1	Yellow	Diffused	2.1	20		
TIL224-2	Yellow	Diffused	6	20		
TIL227-1	Yellow	Clear	6	20	CL-10	High intensity
TIL227-2	Yellow	Clear	15	20	T1-¾	
TIL228-1	Red	Diffused	2.1	20		
TIL228-2	Red	Diffused	6	20		
TIL231-1	Red	Clear	6	20		
TIL231-2	Red	Clear	15	20		
TIL232-1	Green	Diffused	0.5	20	CL-9	
TIL232-2	Green	Diffused	1.3	20	T1-¾	
TIL234-1	Green	Diffused	0.8	20		
TIL234-2	Green	Diffused	2.1	20	CL-10	High Intensity
TIL236-1	Green	Clear	6	20	T1-3¾	
TIL236-2	Green	Clear	15	20		
TIL261 thru TIL270	Red	Diffused	0.5	25		
TIL271 thru TIL280	Green	Diffused	0.8	25	CL-25	One-element thru ten-element arrays
TIL 281 thru TIL290	Yellow	Diffused	0.8	25		

CL-9 and CL-10 are single unit two-leaded packages. CL-25 is a multiple unit assembly. T1 is approximately 0.125 inches in diameter; T1-¾ is approximately 0.2 inches in diameter.

6 APPLICATIONS OF LIGHT-EMITTING DIODES

Logic Level Indicators

In a computer or other device using digital logic circuits, the output of a logic circuit can be monitored with a VLED to give visual indication of its status. If the logic gate output normally cycles between the high and low logic levels, the VLED will produce a dim light under normal cycling of the gate because the VLED is being turned on and off at a rapid rate. If the cycling stops, the light will either go off or produce a bright light indicating that the gate is stopped in either the high or low logic levels. Also, VLEDs can be used to monitor command signals from a computer to various peripheral equipment so the operator knows when a command signal has been sent.

Typical circuits to use a VLED with a TTL integrated circuit are shown in *Figure 6-8*.

Figure 6-8a shows a circuit which produces light when the logic output is high. This circuit has the disadvantage that the diode limits the output voltage to 1.6 volts which is below the "logic one" level. This means that although the circuit makes a useful indicator, it cannot be used to drive other TTL gates.

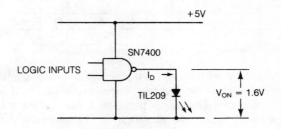

a. VLED on when output is high

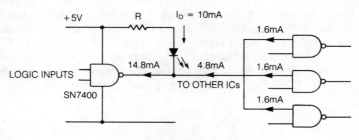

b. VLED on when output is low

Figure 6-8. *Typical Circuits Used to Monitor TTL Logic Gates*

APPLICATIONS OF LIGHT-EMITTING DIODES

6

Figure 6-8b shows a circuit which turns the VLED on when the output of the gate is at a low logic level. The value of R in the circuit determines both the intensity of the light output and the current available to drive other logic gates. For example, if the integrated circuit is a standard SN7400 TTL logic gate, it is capable of sinking 16mA of current and driving up to 10 other TTL gates which require 1.6mA each (fan-out of 10). If a VLED current of 10mA is required to produce the desired light level, then only 16mA-10mA or 6mA are left to sink the inputs of other TTL logic gates. Since standard TTL gates require 1.6mA input current, three gates require 4.8mA and four gates require 6.4mA. However, the driving circuit can sink only 6mA after the VLED uses 10mA, thus, there is not enough current to drive four gates but there is enough to drive three gates. Therefore, the fan-out of the driver (to the other TTL gates) is three. *Table 6-3* lists the values of VLED current, R, and the resulting fan-out for the example circuit.

Table 6-3. *VLED Current, Limiting Resistor (R), and Fan-out for Circuit of Figure 6-8a**

I_{DmA}	0.5	1	2	4	6	8	10	14
R kΩ	6.5	3.2	1.6	0.8	0.51	0.39	0.30	0.22
Fan-Out	9	9	8	7	6	5	3	1

*Assumes TTL Gate with 16mA Current Sink at its Output and 1.6mA Current Source for Each TTL Input

Problems with fan-out of logic gates and loading effects of LEDs are not just the concern of the designer. Careless troubleshooting procedures can also lead to incorrect diagnosis of problems because of the fan-out limitation. Suppose a simple logic probe made of a VLED in series with a 300 ohm limiting resistor (*Figure 6-9a*) is being used to check logic voltage levels in the circuit of *Figure 6-9b*. Referring to *Table 6-3* shows that a TTL gate will be able to provide 10mA of current for the probe and still have a fan-out of three. The probe could be used to check logic levels at points labeled X since each of these points have a fan-out of three or less. However, the probe should not be used to check levels at the point labeled Y because the probe would overload the circuit and possibly change the logic levels of gates to the right of point Y.

It should be pointed out that there are many logic probes available on the test equipment market which use *power from another power supply* to light the test VLED. These probes usually do not effect the fan-out.

6. APPLICATIONS OF LIGHT-EMITTING DIODES

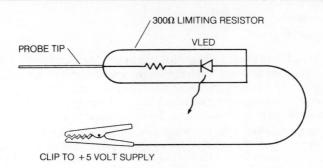

VLED	LOGIC LEVEL
ON	LOW
OFF	HIGH

a. Simple Logic Probe

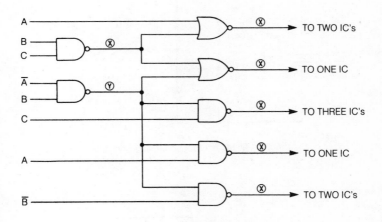

b. Logic Circuit to be Checked

Figure 6-9. Logic Probe Application

APPLICATIONS OF LIGHT-EMITTING DIODES

6

Fault Indicators
Low-Voltage Monitor

Figure 6-10 gives a circuit that can be used to monitor the voltage of an automobile battery and warn the operator when the battery voltage is below 10 volts by turning on the VLED. The important devices in this application are the comparator and the VLED. The comparator is a device which senses two different voltages and provides an output which is either on (low voltage) or off (high voltage) depending on the relative size of the two input voltages. One of the inputs is called the noninverting input, designated by a plus (+) sign, while the other is the inverting input, designated by a minus (−) sign. If V+ is more positive than V− the output is high. If V− is more positive than V+, the output is low. *Figure 6-10b* shows how the output switches with V− set at +5 volts. If V+ is less than +5 volts, the output voltage is low. If V+ is greater than +5 volts, the output voltage is high.

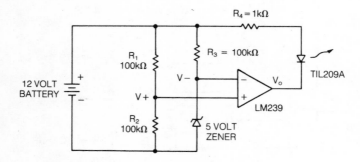

a. Schematic of Circuit for Low Voltage Indicator

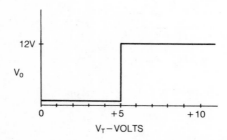

b. Transfer Curve for LM239 with V− equal to +5 volts

Figure 6-10. *Using a VLED for a Low Voltage Fault Indicator*

6. Applications of Light-Emitting Diodes

In the circuit shown in *Figure 6-10a*, the reference voltage at the inverting ($-$) input of the LM239 comparator is set by a 5 volt zener diode and R_3. R_1 and R_2 are used as a voltage divider to provide one-half the battery voltage to the noninverting ($+$) input. When the battery voltage decreases to less than 10 volts, the voltage at the noninverting input goes below 5 volts which is less than the reference voltage. This causes the output of the amplifier to switch from a high level to a low level which turns on the VLED. Current through the VLED is limited by R_4. By using eight such circuits and selecting appropriate values of R_1 and R_2 for each, a battery status display could be made to indicate the battery voltage in one volt steps from 7 volts to 15 volts.

Indicator for Power Failure

Another type of fault indicator can be used to provide information about a fault that may have occurred in the operator's or user's absence. For example, if the power is off in your home for a few minutes, the conventional electric clocks will be slow. Unless you happen to compare their time with another source you might not notice that the clocks were wrong. A fault indicator using a VLED and a latch circuit which requires a reset input from the user could be used to alert you.

Figure 6-11 gives an example of a latch circuit using a relay. When the normally open reset switch is momentarily depressed, the relay is activated. The normally open contacts of the relay close and bypass the reset switch to provide a holding current path for the relay coil when the reset switch is released and its contacts open. The normally closed relay contacts open and interrupt the current to the halfwave rectifier that supplies dc power to the VLED. If the power goes off momentarily, the relay will deactivate which allows the normally closed contacts to close. When the power comes on again, the dc power supply will turn on the VLED to give a visual indication that the power was off at some time. Depressing the reset switch activates the relay to turn off the VLED.

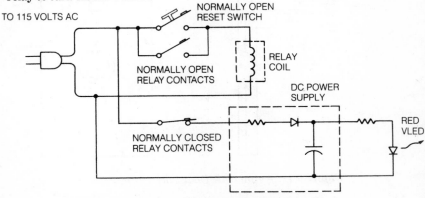

Figure 6-11. Fault Indicator With Relay Latch

Applications of Light-Emitting Diodes — 6

Instrument Panels

Status indicators, such as excessive speed warning, low oil pressure warning, high temperature warning, and seat belt warning, are commonly used in automobile instrument panels to display information to the operator. As discussed in the application example earlier, the major problem is the variation in ambient light. If the indicators are visible in bright sunlight, they may be too bright at night. However, warning indicators are normally off so it is better for them to be too bright at night than not visible during bright sunlight.

In other applications, it is more important for the display to be visible in a wide variety of ambient light conditions without being too bright at night. For example, the mechanical speedometer can be replaced with an array of VLED's which turn on sequentially as speed increases. The brightness of the display can be controlled by a manually operated intensity control, or controlled automatically by using an appropriate light detector to sense ambient light levels and provide a signal to increase or decrease the VLED intensity as needed.

Illumination

An example of using a VLED for illumination is the door lock illumination used on some automobiles. When the door handle is moved, a circuit is activated which turns on an amber VLED and illuminates the lock so that it can be seen in the dark. The mechanical configuration is shown in *Figure 6-12*. The light generated by the VLED moves through the light guide and illuminates the key hole.

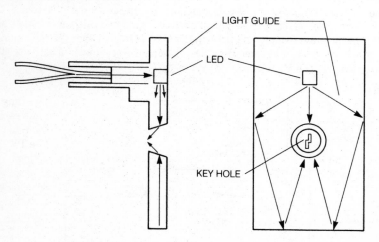

Figure 6-12. *Automobile Door Lock Illuminator*

6 APPLICATIONS OF LIGHT-EMITTING DIODES

Illumination does not need to be visible to be useful. Infrared LEDs are useful in applications where communication from machine to machine is required without distracting people who may be working in the area. For example, infrared (9,300 Å) light sources are used in infrared transmitters and receivers for remote control of television sets as shown in *Figure 6-13*. These units transmit signals by pulsing the infrared LED without interference with the human sensing system and without generating electromagnetic signals which might interfere with the TV or radios nearby. Another advantage of using infrared is that the receiver will not respond to ambient light, jingling keys, or garage door opener transmitters. Since LEDs use low power and they are on only when actually used, a hand-held battery operated transmitter will function for many months before needing a new battery.

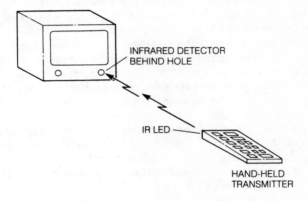

Figure 6-13. Remote Control for TV

Typical circuits for the transmitter and receiver of the remote control unit are shown since all are identical except for the modulating frequency. For the transmitter shown in *Figure 6-13*, a different modulating frequency is used for each control button for a total of 15 frequencies. As each button is pushed, a particular frequency modulates the light produced by the infrared LED in the hand-held transmitter. An infrared detector in the TV detects the transmitted signal and demodulates it; that is, the modulating frequency is removed from the IR light. There is a frequency selective circuit for each of the different modulating frequencies in the remote control receiver. Each one will pass only one frequency so these circuits route the detected frequency to the proper decoder. The decoder determines what function is needed and provides a control signal for the TV circuits to perform that function.

UNDERSTANDING OPTRONICS 6-19

APPLICATIONS OF LIGHT-EMITTING DIODES   6

Ten of the buttons on the remote control transmitter are for the numbers 0 through 9 and are used to key in the TV channel desired. Each number is transmitted as it is keyed, detected at the receiver, and stored in a two-digit register. After the desired channel has been keyed in, the ENTER button on the transmitter is pushed. This signal is also detected and decoded at the receiver. It tells the decoder to use the numbers stored in the register to determine the value of the tuning voltage needed by the electronic tuner to select the desired channel. The tuning voltage is applied to the tuner and the desired channel is immediately selected without sequencing through all channels. The other buttons on the remote control transmitter provide on-off control, up-down volume control, and sound muting.

In Figure 6-14, when switch A on the transmitter is closed, power is supplied to oscillator A which provides a single frequency signal (f_A) to the input of the SN72741 operational amplifier. The amplifier then modulates the light from the infrared LED at that frequency.

The receiver detects the modulated light and provides an input to the two active filters. The top filter is tuned to the frequency (f_A). The signal is amplified by the twin-tee filter and provides an output signal that is used to tune to the channel which has been selected by pushing the channel number buttons on the key pad. The signal is not amplified by the bottom filter since it is tuned to another frequency (f_B).

If switch B on the transmitter is closed, power is supplied to oscillator B which provides a different frequency (f_B) to the SN72741 operational amplifier. This modulates the light at a frequency f_B. The modulated light is detected by the receiver and amplified through the bottom filter. The amplifier output is used to turn the TV off or on.

WHAT HAVE WE LEARNED?

1) Light from a VLED is directly proportional to current.
2) The eye does not detect changes in light intensity if the ratio is less than 2 to 1. Therefore, small changes in VLED current do not produce noticeable changes in light.
3) Viewing angle is determined both by the lens design and the choice of clear or diffused lens bodies.
4) Diffused lenses produce soft light with a wide viewing angle whereas clear lenses produce a higher intensity light with a narrower viewing angle.
5) An electronic circuit must be able to provide on the order of 0.5mA to 20mA of current at 1.6 to 2 volts to turn on a VLED. (0.8 to 40 milliwatts)
6) Filters can be used to improve contrast ratio.
7) VLEDs are suitable in many applications because of high reliability, fast response and low cost.
8) The brightness of an VLED may be controlled by turning it on and off at a rapid rate because of the averaging effect of the human eye.

6 APPLICATIONS OF LIGHT-EMITTING DIODES

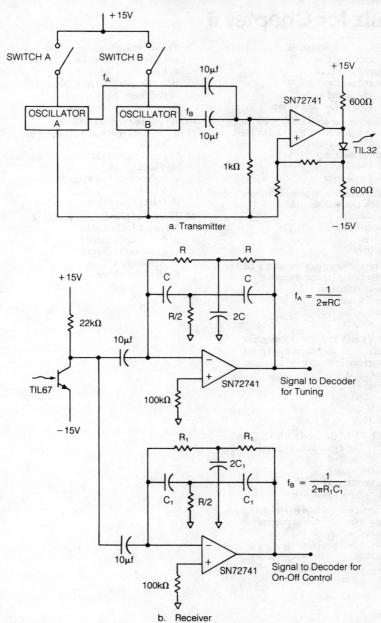

Figure 6-14. Optotransmitter and Receiver

APPLICATIONS OF LIGHT-EMITTING DIODES 6

Quiz for Chapter 6

1. The most common colors available from VLEDS are
   a. red and blue
   b. red and white
   c. green and amber
   d. red, yellow and green

2. In order for an VLED to be easily seen the current must be at least
   a. 10 microamps
   b. 0.1 milliamp
   c. 10 milliamps
   d. 1 milliamp

3. If a VLED is to be visible from a side view the lens should be
   a. clear
   b. diffused
   c. red
   d. green

4. The voltage drop across a VLED that is "on" is approximately
   a. 10 millivolts
   b. 0.6 volt
   c. 2 volts
   d. 5 volts

5. A VLED with a wavelength of _____ angstroms is a good warning indicator for a human operator.
   a. 4500
   b. 5500
   c. 6500
   d. 7500

6. If a VLED is used with a SN7400 TTL logic gate, the fanout is
   a. decreased
   b. increased
   c. unchanged
   d. none of the above

7. Power dissipation in an LED is an important electrical parameter because
   a. it affects reliability.
   b. it determines temperature.
   c. it affects battery life.
   d. all of the above.

8. Temperature is an important consideration because
   a. it affects reliability.
   b. it changes light intensity.
   c. it changes light wavelengths.
   d. all of the above.

9. A red optical filter will
   a. improve contrast ratio for a red VLED.
   b. improve contrast ratio for a green VLED.
   c. reduce contrast ratio for a red VLED.
   d. none of the above.

10. In most applications it is best to drive a VLED from a
    a. dc voltage source.
    b. dc current source.
    c. ac voltage source.
    d. all of the above.

7 APPLICATIONS OF PHOTODETECTORS

Applications of Photodetectors

ABOUT THIS CHAPTER

In the last chapter, applications of optically-coupled systems were discussed with emphasis on the light-emitting diodes as the source. In this chapter the detector used in such systems is the subject of concern. One of the most useful types of photodetectors is the PN junction diode, which is referred to as a photodiode when made for use as a photodetector.

WHAT ARE THE CHARACTERISTICS OF PHOTODIODES?

Electro-optical Characteristics of Photodiodes

As mentioned in Chapter 3, light falling on a suitably doped semiconductor PN junction will produce hole-electron pairs (*Figure 3-4*). These generated carriers produce a current flow in an external circuit connected to the PN junction. This current is called the photocurrent.

The ability of a generated hole-electron pair to contribute to the photocurrent depends on the hole and electron being rapidly separated from each other before they can bump together and cancel out. When a PN junction is reverse biased, a depletion region forms at the junction of the P and N materials. If the PN junction is thought of as a capacitor with the depletion region as the dielectric between the charged plates, as shown in *Figure 7-1a*, then it is easy to see that a hole and electron created in the depletion region would be rapidly pulled apart by the oppositely charged plates. Thus, the effectiveness of a PN junction diode as a photodetector can be enhanced by widening the depletion region to give more opportunity for hole-electron pairs to form there.

There are two ways to increase the width of the depletion region. One way is to include a layer of pure semiconductor material (called intrinsic material) between the P and N type materials. The resulting diode is called a PIN diode and a cross section is shown in *Figure 7-1b*. The name PIN emphasizes the layer of *intrinsic* material sandwiched between the P and N type materials. A diode is represented schematically as shown in *Figure 7-1c* when it is used as a photodiode.

APPLICATIONS OF PHOTODETECTORS 7

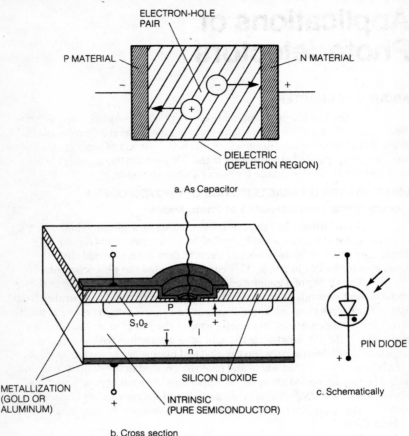

Figure 7-1. PIN Diode with Reverse Bias

If the reverse bias (positive on N material, negative on the P material) is increased across the PN junction, then the depletion region widens. This gives a second way to widen the depletion region to enhance the diode's use as a photodetector. Since the leakage current increases with reverse bias voltage, it is desirable to use as small a reverse bias as possible. Therefore, these two conditions are counterproductive and the final operating conditions are a compromise. Applications that have high intensity light sources can tolerate much more steady-state detector current than systems that have sources with very low low light intensity. The intended use of the photodiode will dictate how it will be operated; that is, whether it has a reverse bias and how much, or whether it has any external bias.

7 APPLICATIONS OF PHOTODETECTORS

Some of the most important photodiode characteristics concern the relationship between the photocurrent and the light falling on the photodiode. Some of these characteristics have been defined and mentioned previously, but let's examine these characteristics in more detail and determine some typical values for specific devices.

Quantum Efficiency

At the quantum level, the important factor for a semiconductor photodetector is the number of hole-electron pairs generated per photon of light falling on the PN junction. This has been defined previously as the quantum efficiency. If it is included as a line item in a data sheet, it is given at a specified wavelength. It also may be given in graph form and can be evaluated at a selected wavelength. For example, *Figure 7-2a* specifies the quantum efficiency of a typical PIN photodiode as 0.75 electrons/photon at 7,700Å.

Flux Responsivity

Another important characteristic is the flux responsivity. This quantity is the ratio of the output photocurrent in amperes to the watts of radiant flux causing the photocurrent. It too, may be given as a line item at a specific wavelength or in graph form. If in graph form, a specific value at a specific wavelength can be evaluated. For example, *Figure 7-2a* has the flux responsivity specified as 0.45 amperes/watt at 8,000Å for a typical PIN photodiode.

Spectral Response

Each of these characteristics can be plotted against wavelength to provide an important specification for a photodiode—its spectral response. The spectral response gives the range of light frequencies to which the photodiode will respond and also shows the relative size of the response (i.e., the generated photocurrent) at each wavelength. The spectral response can be given in quantum efficiency as shown in *Figure 7-2a* or it can be given as just a relative response as shown in *Figure 7-2b*.

Photocurrent vs Irradiance

Photocurrent can also be given as a graph showing the photocurrent as a function of the irradiance (incident radiation) in watts per square meter. This is shown in *Figure 7-3*. Since photocurrent varies with the reverse bias applied to the diode, photocurrent may be graphed as a function of the reverse bias at particular levels of irradiance as shown in *Figure 7-4*. In this case, the irradiance is in watts per square meter. This graph clearly shows the advantages of operating with reverse bias. It is characterized at a given wavelength.

APPLICATIONS OF PHOTODETECTORS 7

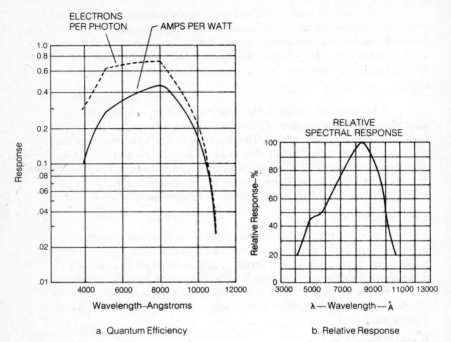

Figure 7-2. Spectral Response may be Given as a Quantum Efficiency or as a Relative Response

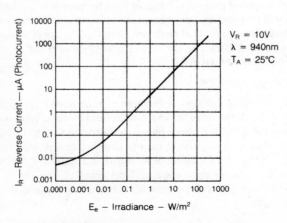

Figure 7-3. Photocurrent Versus Irradiance

7 APPLICATIONS OF PHOTODETECTORS

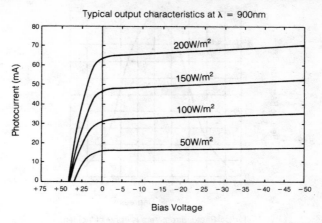

Figure 7-4. Photocurrent Versus Bias Voltage at Various Light Levels

Incidence Response

In some applications, the photodiode is illuminated uniformly over the entire surface. A measure of the response of a photodetector when it is uniformly illuminated over its entire surface is the incidence response. The incidence response is the ratio of the photocurrent in amperes to the incident radiation in watts per square centimeter. This characteristic for a typical photodiode is 0.001 amps/watt/cm^2; that is, 1 milliampere of photocurrent for every watt per square centimeter of radiation. Incidence response is usually specified only at a single wavelength.

Optical Characteristics of Photodiodes

The position of the photodetector relative to the intended light source or sources, or relative to unwanted sources, is important to many applications. The directional sensitivity of the photodiode is used in this case. This characteristic shows the relative reponse of the photodiode as a function of the angle of incidence of incoming light. It gives an indication of which areas of the photodiode are the most sensitive to light. In applications, it gives an indication of how critical the alignment between source and detector must be, or how much unwanted sources will interfere with detection of light from the wanted source. The data may be plotted as photocurrent versus angle of incidence as shown in *Figure 7-5a* or it can be plotted in polar coordinates on a plane which cuts through some portion of the photodiode package as shown in *Figure 7-5b*. Notice that a domed lens causes a narrower field of view than a flat lens on the same device.

APPLICATIONS OF PHOTODETECTORS 7

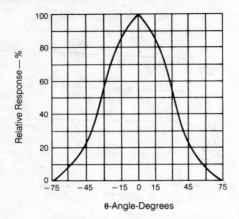

a. Typical Photocurrent Versus Angle of Incidence

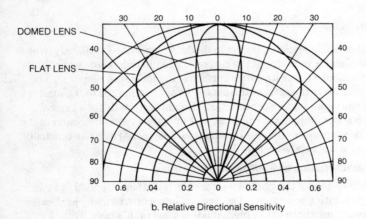

b. Relative Directional Sensitivity

Figure 7-5. Two Often Used Techniques of Displaying Photocurrent Response Versus Angle of Incidence (Directional Sensitivity)

Electrical Characteristics of Photodiodes

The electrical characteristics of a photodiode are similar to the characteristics of any semiconductor diode. *Figure 7-6* shows a typical set of specifications from a photodiode data sheet.

7 APPLICATIONS OF PHOTODETECTORS

electrical characteristics at 25°C free-air temperature

	PARAMETER		TEST CONDITIONS		MIN	TYP	MAX	UNIT
A	$V_{(BR)}$	Breakdown Voltage	$I_R = 100~\mu A$,	$E_e^\dagger = 0$	30			V
B	I_D	Dark Current	$V_R = 10~V$,	$E_e^\dagger = 0$		5	50	nA
C	I_L	Light Current	$V_R = 10~V$,	$E_e^\dagger = 250~\mu W/cm^2$ at 940 nm	10	15		μA
D	C_T	Total Capacitance	$V_R = 3~V$,	$E_e^\dagger = 0$, f = 1mHz		35	50	pF
E	t_r	Rise Time	$V_R = 10~V$,	$R_L = 1~k\Omega$		100		ns
	t_f	Fall Time	$V_F = 10~V$,	$R_L = 1~k\Omega$		100		ns

†Irradiance (E_e) is the radiant power per unit area incident on a surface.

***Figure 7-6.** Characteristics of TIL100 Photodiode*

V_{BR}

One of these specifications is the maximum reverse bias voltage that can be applied without damaging the diode. This is called the breakdown voltage (V_{BR}) of the diode. For the TIL100, V_{BR} has a minimum value of 30 volts as shown in *Figure 7-6*, line A. The normal breakdown voltage for PIN diodes is 30 to 50 volts, while avalanche photodiodes can have reverse voltage specifications of 150-200 volts.

I_D and I_L

Another important specification of a reverse biased diode is the reverse saturation current. For a photodiode, this current is referred to as the dark current (I_D) because it is the current that flows even though no light is applied to the junction. The dark current must be small compared with the photocurrent from the diode. If necessary, the dark current can be reduced by lowering the reverse bias voltage applied to the diode. The dark current for the TIL100 has a typical value of 5 nanoamps and a maximum value of 50 nanoamps as shown in *Figure 7-6*, line B. The differential between the dark current and the photocurrent with a given intensity of a light, such as I_L in line C of *Figure 7-6*, will determine design margins for the detector in optically-coupled systems. Dark currents range from picoamperes to nanoamperes.

As previously stated, dark current is the leakage current that flows when no light falls on the junction. Leakage current in PN junction devices is very surface dependent. Sometimes a P-type diffusion made into N-type substrate will cause the N-type to invert to P-type in regions where no diffusion was made. This causes high leakage current and thus, high values of dark current.

APPLICATIONS OF PHOTODETECTORS

7

In some optoelectronic devices and, in particular, high-resistivity PN junction devices, an extra diffusion of N^+ (heavily-doped N material) is made between the PN junction and the edge of the silicon substrate. This is called a *guard ring*. It guards against inversion layers going to the edge and causing large leakage currents. Sometimes the guard ring is connected to an external lead on the package so that it can be tested. However, for optical couplers the guard ring must be connected to a point that keeps it at a low potential with respect to the phototransistor elements. Otherwise it can form a plate of a capacitor with the light source and cause inversion layers at the surface of the phototransistor if the isolation voltage is high.

C_T

A third important specification of any diode is the capacitance of the PN junction. (See *Figure 7-6*, line D.) This capacitance limits how quickly a photodiode can respond to changes in the light shining on the junction. Depending on the junction area, this capacitance varies from a picofarad to tens of picofarads. The larger this capacitance, the more slowly the photodiode responds to light changes. The capacitance can be reduced by operating with a higher reverse voltage as shown in *Figure 7-7*; however, a higher dark current may result from the higher reverse voltage. *Figure 7-7* is for a large junction area photodiode (TIL100).

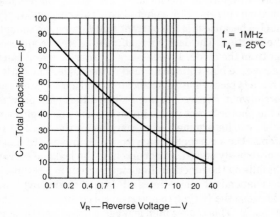

Figure 7-7. *Total Capacitance Versus Reverse Voltage for the TIL100 Silicon Photodiode*

7. APPLICATIONS OF PHOTODETECTORS

t_r, t_f and f_c

The frequency response of the photodiode is also important. As mentioned earlier, the photodiode frequency response must not be confused with the photodiode spectral response. The frequency response refers to how quickly the photocurrent can respond to changes in the radiation incident on the photodiode. As previously mentioned, frequency response can be expressed in two ways: by switching characteristics or by cutoff frequency.

In *Figure 7-6* line E, the rise and fall time of the TIL100 photodiode are specified to indicate the speed of response of the diode to pulses of light radiation. A switching time circuit is shown in *Figure 7-8*. A light-emitting source is pulsed on and off to produce pulses of light current from the detector which produce voltage pulses across the load resistor, R_L. These voltage pulses are measured with an oscilloscope and the rise and fall time

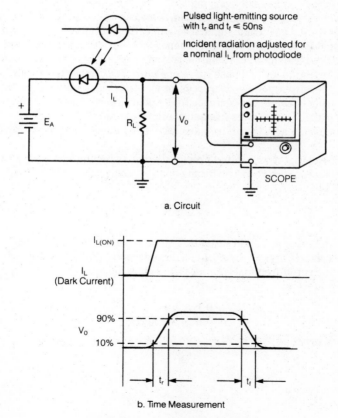

Figure 7-8. Switching Time Circuit

APPLICATIONS OF PHOTODETECTORS

7

determined as shown. The ON light current is adjusted to a nominal value consistent with the specifications for the photodiode by adjusting the light radiating from the light-emitting source. For example, line C of *Figure 7-6* gives the light current specification for the TIL100. To measure switching time for the TIL100, the nominal light current might be 5 to 8 microamperes. Similarly, the voltage applied as reverse voltage in the switching circuit is that typically used in an application and is well below the minimum breakdown voltage (Line A of *Figure 7-6*). All of the switching time measurements are with the radiating source at one specific wavelength.

The second way to express frequency response, the photodiode cutoff frequency, gives an indication of how the photodiode responds to continuously varying signals, not just step changes. For this measurement the incident radiation is varied (modulated) continuously in the same type of circuit as the switching time circuit. This is shown in *Figure 7-9a*. The amplitude of the continuously varying output voltage produced by the continuously varying input light radiation is measured across R_L. The input radiation is set at a particular signal frequency, the output voltage measured, then the input is changed to a new frequency and the output voltage measured again, etc. The incident input radiation is kept constant for each frequency. As shown by the frequency response curve of *Figure 7-9b*, as the frequency is increased, the output voltage decreases for the same constant input radiation because the photo current is decreased. The point where the photocurrent has decreased to 0.707 of its lower frequency relative value of 1.0 is defined as the cutoff frequency. Cutoff frequencies of photodiodes range from 20kHz to 2 GHz.

Recall that it was stated previously that switching times can give an indication of frequency response by an equation which was given in Chapter 4. It is

$$f_c = \frac{0.35}{t_r}$$

where f_c is in hertz and t_r is in seconds. The rise time of the TIL100 was 100 nanoseconds (*Figure 7-6*) or 0.1 of a microsecond (0.1×10^{-6} second). Therefore the cutoff frequency for a TIL100 photodiode is:

$$f_c = \frac{0.35}{0.1 \times 10^{-6}}$$
$$= 3.5 \times 10^6 \text{ hertz}$$

or 3.5 megahertz.

APPLICATIONS OF PHOTODETECTORS

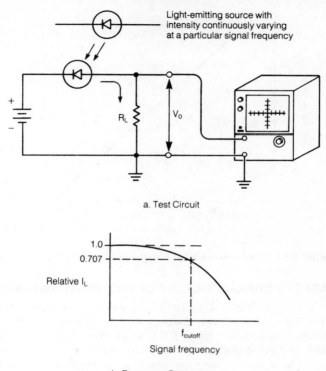

Figure 7-9. Measuring Frequency Response

Figure 7-10 shows how the cutoff frequency varies with different values of load resistor in the measuring circuit. As load resistance increases, cutoff frequency decreases. This should be expected since the cutoff frequency is inversely proportional (it goes down as something else goes up) to the load resistance multiplied by the capacitance of the photodiode.

P_D

Semiconductor junctions are very sensitive to temperature. Power dissipation in a device (voltage across device times the current through the device) determines how much the temperature of the semiconductor material will rise. Particular attention must be paid to the power dissipation ratings of the devices so that the junction temperature is not exceeded.

APPLICATIONS OF PHOTODETECTORS

7

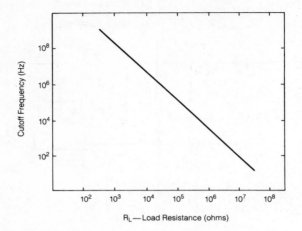

Figure 7-10. Photodiode Cutoff Frequency Versus Load Resistance

WHAT ARE THE CHARACTERISTICS OF PHOTOTRANSISTORS?

In many applications, photodiodes are used in conjunction with an amplifier to increase the effect of the photocurrent. Since a conventional transistor contains a reverse biased PN junction (the collector-base junction) and has the ability to amplify a current applied to the base of the transistor, it would appear to have everything necessary for a photodiode and amplifier in one package. This is indeed the case and the phototransistor is a very useful photodetector. A cross section of a typical phototransistor is shown in *Figure 7-11a*. Its schematic symbol is shown in *Figure 7-11b*.

In a phototransistor, light falling on the collector base junction results in a base photocurrent. This base photocurrent is multiplied by the forward current transfer ratio h_{FE} (the ratio of the collector current to the base current) of the phototransistor to produce the collector photocurrent. In many phototransistors, the base lead is not made available externally. However, in some applications, it may be desirable to use a phototransistor with a base lead to allow additional control over the phototransistor.

7 APPLICATIONS OF PHOTODETECTORS

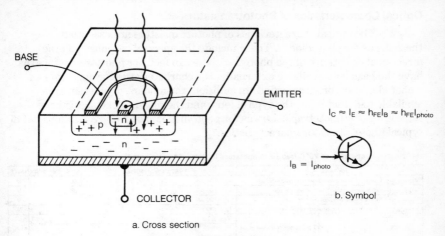

Figure 7-11. Typical Phototransistor

Electro-optical Characteristics of Phototransistors

As mentioned previously, the ratio of the photocurrent, I_L, to the dark current I_D, sets the design margins for the sensitivity of the optically-coupled system. It would seem that, since I_L is increased for phototransistors because of the built-in gain, the I_L over I_D ratio would increase. However, the phototransistor produces no improvement in the photocurrent-to-dark current ratio over the photodiode because the reverse saturation current I_{CBO} which forms the dark current also is amplified by the forward current transfer ratio of the transistor.

The collector to base junction of a phototransistor is much larger in area than the anode-cathode junction of a diode: Because of the large capacitance formed by the large area base-collector junction, the phototransistor has a poorer frequency response than a photodiode. Another limitation of the phototransistor is that the region over which the phototransistor characteristics are linear is more limited than the corresponding operating region for a photodiode. Because of these limitations of a phototransistor as compared to a photodiode, the particular application must be carefully considered in order for one to decide whether to use a phototransistor or a photodiode with a separate amplifier.

APPLICATIONS OF PHOTODETECTORS

7

Optical Characteristics of Phototransistors

The optical characteristics of phototransistors are essentially the same as for photodiodes. These include the spectral response and the directional sensitivity of the phototransistor. In fact, for phototransistors that have the base lead available externally, the characteristics for operation as a photodiode are often specified. In addition, the amount of junction area available to detect light, the type of lens, and the physical packaging have the same effect for phototransistors as for photodiodes. See *Figure 7-12* for a set of typical phototransistor characteristics.

electrical characteristics at 25°C free-air temperature (unless otherwise noted)

PARAMETER		TEST CONDITIONS			MIN	TYP	MAX	UNIT
$V_{(BR)CBO}$	Collector-Base Breakdown Voltage	$I_C = 100\ \mu A$,	$I_E = 0$,	$E_e = 0$	50			V
$V_{(BR)CEO}$	Collector-Emitter Breakdown Voltage	$I_C = 100\ \mu A$,	$I_B = 0$,	$E_e = 0$	30			V
$V_{(BR)EBO}$	Emitter-Base Breakdown Voltage	$I_E = 100\ \mu A$,	$I_C = 0$,	$E_e = 0$	7			V
$V_{(BR)ECO}$	Emitter-Collector Breakdown Voltage	$I_E = 100\ \mu A$,	$I_B = 0$,	$E_e = 0$	7			V
I_D	Dark Current	Phototransistor Operation	$V_{CE} = 10\ V$,	$I_B = 0$,	$E_e = 0$			0.1	μA
			$V_{CE} = 10\ V$, $T_A = 100°C$	$I_B = 0$,	$E_e = 0$,		20		μA
		Photodiode Operation	$V_{CB} = 10\ V$,	$I_E = 0$,	$E_e = 0$			0.01	μA
I_L	Light Current	Phototransistor Operation	$V_{CE} = 5\ V$, See Note 2	$I_B = 0$,	$E_e = 5\ mW/cm^2$,	5	22		mA
		Photodiode Operation	$V_{CB} = 0$ to $50\ V$, See Note 2	$I_E = 0$,	$E_e = 20\ mW/cm^2$		170		μA
h_{FE}	Static Forward Current Transfer Ratio		$V_{CE} = 5\ V$,	$I_C = 1\ mA$,	$E_e = 0$		200		
$V_{CE(sat)}$	Collector-Emitter Saturation Voltage		$I_C = 2\ mA$, See Note 2	$I_B = 0$,	$E_e = 20\ mW/cm^2$,		0.2		V

NOTE 2: Irradiance (E_e) is the radiant power per unit area incident upon a surface. For these measurements the source is an unfiltered tungsten linear-filament lamp operating at a color temperature of 2870°K.

switching characteristics at 25°C free-air temperature

PARAMETER		TEST CONDITIONS	TYPICAL	UNIT
t_r	Rise Time	Phototransistor Operation	$V_{CC} = 5\ V$, $I_L = 800\ \mu A$, $R_L = 100\ \Omega$, See Test Circuit A	8	μs
t_f	Fall Time			6	
t_r	Rise Time	Photodiode Operation	$V_{CC} = 0$ to $50\ V$, $I_L = 60\ \mu A$, $R_L = 100\ \Omega$, See Test Circuit B	350	ns
t_f	Fall Time			500	

PARAMETER MEASUREMENT INFORMATION

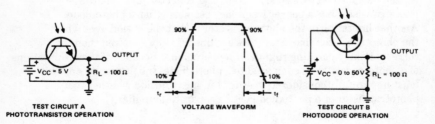

Figure 7-12. Typical Phototransistor Data Sheet Specifications

7 APPLICATIONS OF PHOTODETECTORS

Electrical Characteristics of Phototransistors

The dark current for a phototransistor is the reverse saturation current which flows in the reverse biased collector-base junction with the emitter lead open-circuited. This current is designated I_{CBO} where the subscript CBO refers to the collector-base junction with the emitter lead open-circuited. As mentioned above, this current also is amplified by the transistor. Usually this amplified dark current in the collector is given. This current is designated as I_{CEO} where the subscript CEO refers to the collector-emitter current with the base open-circuited.

Since there are three terminals to which voltages can be applied in a phototransistor, several breakdown voltages can be specified. Those usually specified are BV_{CBO}, the collector-to-base breakdown voltage with the emitter open-circuited; BV_{CEO}, the collector-to-emitter breakdown voltage with the base open-circuited; and BV_{EBO}, the emitter-to-base breakdown voltage with the collector open-circuited.

Other characteristics of interest that are usually given in data sheets are the: maximum collector current, maximum power dissipation, current amplification factor h_{FE} (sometimes called the amplification factor, β), and voltage between the collector and emitter when the transistor is saturated (switched on) $V_{CE(SAT)}$.

Since phototransistors are often used in switching applications in which the phototransistor is switched between cutoff and saturation, it is desirable that $V_{CE(SAT)}$ be as small as possible.

The phototransistor can be operated as a photodiode by leaving the emitter lead open; therefore, characteristics which relate the radiation incident in the collector base junction to the base and collector photocurrents (photodiode operation) are also very important. In fact, many data sheets show both photodiode and phototransistor operation curves for phototransistors. These measures of photocurrent versus incident radiation are the collector-to-emitter radiation sensitivity (S_{RCEO}) as shown in *Figure 7-13a* and the collector-to-base radiation sensitivity (S_{RCBO}) as shown in *Figure 7-13b*. S_{RCEO} is the ratio of collector photocurrent to incident irradiance with the base open-circuited. S_{RCBO} is the ratio of collector photocurrent to the incident irradiance with the emitter open-circuited. The phototransistor gain, β can be calculated from $\beta = \frac{S_{RCEO}}{S_{RCBO}}$. For example in *Figure 7-13a*, for E_e equal 10mW/cm^2 and $V_{CE} = 1V$, the collector current is 12mA. In *Figure 7-13b*, at the same irradiance E_e, the collector current is 0.08mA.

The transistor gain β is:

$$\beta = \frac{12 \times 10^{-3}}{0.08 \times 10^{-3}}$$
$$= 150$$

APPLICATIONS OF PHOTODETECTORS 7

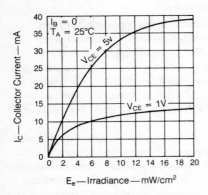

a. Collector-to-Emitter Radiation Sensitivity (Phototransistor Operation)

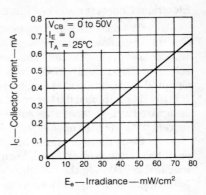

b. Collector-to-Base Radiation Sensitivity (Photodiode Operation)

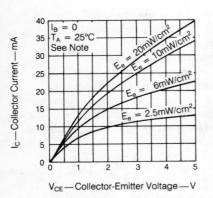

c. Collector Photocurrent as Function of Incident Light

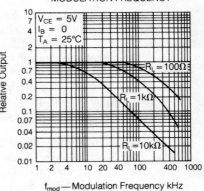

d. Frequency Response Curve

NOTE: Irradiance (E_e) is the radiant power per unit area incident upon a surface. For these measurements the source is an unfiltered tungsten linear-filament lamp operating at a color temperature of 2870°K.

***Figure 7-13.** Graphical Data for a Typical Phototransistor Provides Information for a Wide Range of Applications*

7 APPLICATIONS OF PHOTODETECTORS

The collector photocurrent also may be given in graphs of collector-to-emitter photocurrent as a function of incident flux density and supply voltage to the phototransistor as shown in *Figure 7-13c*. If the base contact is available, the base photocurrent also may be graphed as a function of incident flux density.

As with photodiodes, the frequency response of a phototransistor is also of interest. These characteristics, like the photodiode, can either be given in terms of switching times or in terms of cutoff frequency. A typical frequency response curve is shown in *Figure 7-13d*.

Like photodiodes, phototransistor parameters are temperature dependent; therefore, maximum power dissipation must be calculated for the application to make sure manufacturer's maximum ratings for temperature are not exceeded. In some applications, special heat sinks or equipment cooling may be required to maintain a device within data sheet ratings.

Other Photodetectors

Some circuits may require the use of some other type of photodetector for special requirements. The photo field-effect transistor, avalanche photodiode, and photothyristor are examples of special photodetectors.

Photo Field-Effect Transistor

Getting enough photocurrent output may be the specific requirement for an optically coupled system that must be solved by either using a phototransistor or a photodiode and separate amplifier. However, if space is limited and superior frequency response is required, the photo field-effect transistor may be a better choice.

A cross section of a field-effect transistor is shown in *Figure 7-14*. It is made with a semiconductor technology called MOS which stands for Metal-Oxide-Semiconductor. The transistor is made from a sandwich of semiconductor material, on which is built a layer of insulating material called silicon dioxide, on which is deposited a layer of aluminum metal. The semiconductor portion of the transistor is a piece of silicon consisting of parts that are N-type and parts that are P-type. In *Figure 7-14* the ends of the material are N-type and the middle is P-type. No electrical contact exists between the metal plate and the silicon material so no electricity passes through this layer.

When +10 volts and ground, are applied to the end N-type pieces, as shown in *Figure 7-14*, and the metal plate is held at ground, no current flows between the end pieces because it is blocked by the N P junction nearest the +10 volt connection. The transistor is in the OFF state.

APPLICATIONS OF PHOTODETECTORS

7

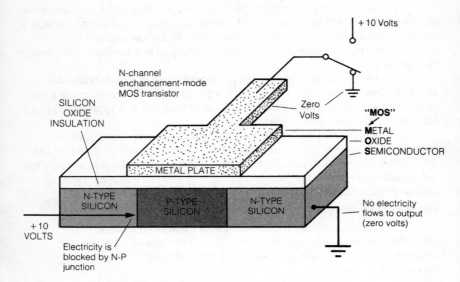

Figure 7-14. *Field-effect Transistor (OFF State)*

Figure 7-15 shows what happens when +10 volts is applied to the metal plate. Electrons are drawn out of the plate creating a positive charge over the surface of the plate. The electric field created by this charge extends down through the oxide layer and *turns* the P-type material into N-type as shown. Thus, a channel of N-type material is formed to bridge the gap between the N-type end pieces and current flows. Now the transistor is in the ON state. It will remain this way as long as the voltage is maintained in the metal plate. Because an N-type channel has been formed to turn on the transistor, this device is called a N-channel enhancement mode MOS field-effect transistor. It is called field-effect because the field through the oxide is what controls the current flow. The control of current need not be digital (from ON to OFF as has been previously described); it can be very gradual. For this reason, field-effect transistors can be used just as bipolar transistors for linear amplification.

Because semiconductor materials, depletion layers and fields are associated with field-effect transistors, they also can be used as photodetectors.

7 APPLICATIONS OF PHOTODETECTORS

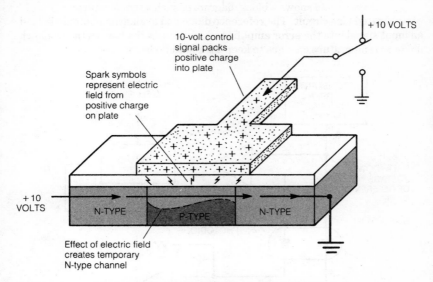

Figure 7-15. Field-effect Transistor (ON State)

In some applications where a phototransistor may not be adequate, a photo field-effect transistor may provide the additional performance needed to satisfy the application requirements so that a separate diode and amplifier combination will not have to be used. A photo field-effect transistor can provide all the desirable characteristics of a phototransistor plus higher gain and superior frequency response. However, the photo field-effect transistors suffer from limited linear range and have higher dark currents than photodiodes. This can limit the design margin for systems that have limited sensitivity.

Avalanche Photodiode and Reference Diode

Avalanche photodiodes were discussed in Chapter 3. Because this diode is operated in the avalanche region, one photon produces an avalanche of energetic electron-hole pairs that produce further electron-hole pairs; therefore, an amplification of photocurrent results. Avalanche photodiodes are extremely temperature sensitive and their gain is very dependent on reverse bias; however, they are important devices because they provide very high-speed response for the detection of light with frequencies that are near-infrared and visible light.

In order to use these devices more effectively as photodetectors, a combination of an avalanche diode and a reference diode are matched together in a package so that they can be used in a temperature compensated bias circuit that holds avalanche gain constant for large changes in temperature.

APPLICATIONS OF PHOTODETECTORS

7

Figure 7-16 shows a block diagram of such a temperature compensated bias circuit. The reference diode and avalanche photodiode feed an input signal into the error amplifier which adjusts the bias on the avalanche diode as temperature changes to keep the photodiode gain constant.

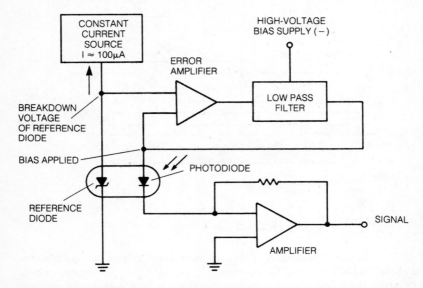

Figure 7-16. *Block Diagram of Temperature Compensating Bias Circuit*

Photothyristors

A thyristor is a semiconductor switching device that is turned on by a control current, but doesn't require any control current once it is turned on. A common type of thyristor is the silicon controlled rectifier (SCR). SCRs are normally used to handle very large currents required for switching motors, generators, or power distribution lines. If an optically coupled system output is required to supply such large currents, it may be possible to use a special optically controlled SCR called a photothyristor.

Figure 7-17a shows the semiconductor structure of an SCR. It is a device with PNPN type semiconductor material layered one on top of the other. To make it easier to understand how it operates, it can be thought of as two transistors structured together as shown in *Figure 7-17b*. The three regions on the left form a PNP transistor and the three regions on the right form an NPN transistor.

7 APPLICATIONS OF PHOTODETECTORS

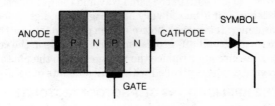

a. Basic Structure

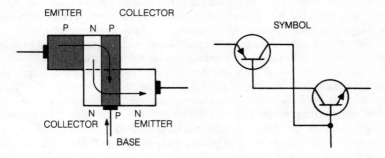

b. Two-transistor Stack

Figure 7-17. Semiconductor SCR

If a plus voltage is applied to the P type material on the left and ground to the N material on the right, current could flow through the first PN junction as it does in any forward biased diode. However, it is blocked from flowing by the next PN junction unless the gate terminal is positive with respect to ground (more than the silicon threshold voltage of 0.7V). In other words, the NPN material stack that forms the NPN transistor acts the same as an NPN transistor in the OFF state because no base current is flowing.

Once the voltage on the gate terminal is above the threshold and current is pumped into the base of the NPN transistor, its gain causes a much higher collector current to flow. This NPN collector current is base current for the PNP transistor which, through gain, causes collector-to-emitter current to flow. The PNP collector-to-emitter current reinforces the original NPN base current that initially triggered the unit and the forward bias current is maintained even when the gate voltage is removed. Thus, for a small initial trigger current, a very large through current flows. The only way to stop the large current flow is to break the circuit or reduce the voltage across the unit to near zero.

APPLICATIONS OF PHOTODETECTORS

Photo SCRs are built in the same way and the gate terminal is available for biasing. If it is biased very near triggering, light energy in the form of photons generates enough hole-electron pairs to trigger the unit into conduction. To turn the unit off, provisions must be made in the control circuits to break the circuit or reduce the voltage across the photo SCR. Thus very large currents can be controlled with small amounts of light energy.

WHAT ARE THE APPLICATIONS OF PHOTODETECTORS?
Photographic Light Meter

Many applications of photodetectors do not require a dedicated light source, but only require that changes in ambient light be detected. As discussed previously, a photocurrent is produced in a junction photodetector which is proportional to the incident radiation. If we operate a junction photodetector in a linear region of its photocurrent characteristic and provide some external circuitry to use the changes in photocurrent to drive a calibrated display, we will have an effective light meter. The block diagram is shown in *Figure 7-18*. The junction photodetector spectral response characteristic would be selected to ensure the greatest sensitivity at the particular light frequencies to be used. For instance, if the light is to expose photographic film, the light meter should be optimized and calibrated for the film's sensitive frequencies since the spectral response of film is different than the spectral response of the eye.

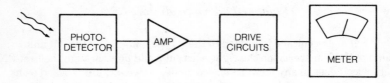

Figure 7-18. Photographic Light Meter

Light meters that indicate actual light output have meter scales calibrated in foot-candela or foot-lamberts. However, photographic light meters for film exposure may have only color scales; one for f stop and one for shutter speed. When the film speed is set by a front dial on the light meter, the meter indicates the f stop and shutter speed that is to be set on the camera.

7 APPLICATIONS OF PHOTODETECTORS

Photographic Flash Control

Another important use of junction photodetectors is in automatic flash units for cameras. The flash is turned on by switch contacts synchronized with the shutter opening and junction photodetectors detect reflected light from the flash. When enough light has been detected to ensure proper exposure, the automatic flash control circuit turns off the flash unit.

The flash is caused by the stored charge on a large capacitor draining off through the flash tube. When the photodetector has detected enough reflected light for proper exposure, the flow of current through the flash tube must be interrupted. As shown in *Figure 7-19a*, one way to interrupt the current is to close a switch in parallel with the flash tube. When the switch is closed, all the remaining charge on the capacitor will be shunted around the flash tube. A disadvantage of this method is that it drains all the charge off the storage capacitor for each flash even if all the charge is not needed for the flash. This is a needless waste of energy and, if the unit is battery operated, requires that the battery be replaced or recharged more frequently. Also, the same amount of time is required to recharge the capacitor regardless of how little charge was actually needed for the flash.

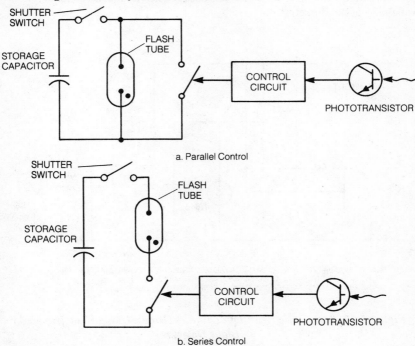

Figure 7-19. Control of a Flash Tube

APPLICATIONS OF PHOTODETECTORS 7

An alternative method to interrupt the current is to place the control switch in series with the flash tube and storage capacitor as shown in *Figure 7-19b*. In this case, when the photodetector senses the proper amount of exposure, the switch is opened to stop the current flow through the flash tube, but the remaining capacitor charge is not drained off.

In this application, the photodetector needs to have a spectral sensitivity compatible with the spectrum produced by the flash tube. The photodetector should also have high directional sensitivity to ensure that the light detected for control of the flash tube will be primarily from the scene being photographed.

Headlight Dimmer

An interesting use of a phototransistor is in the automatic headlight dimmer circuit shown in *Figure 7-20*. The photocurrent generated by the phototransistor is used to control the on-off state of a driver transistor which, in turn, controls the relay that switches between high and low beams. Since the detector needs to sense the light emitted by on-coming headlights, it needs to be very sensitive and directional. A lens which focuses light onto the phototransistor can be used to increase both sensitivity and directional capability of the detector.

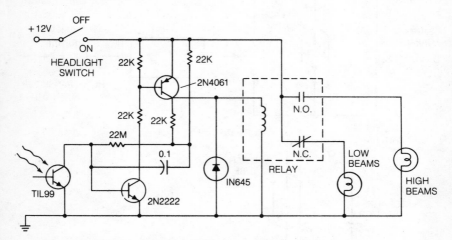

RELAY: 12V, 0.3A COIL: 20A, FORM C CONTACTS OR SOLID-STATE SWITCHING OF 16A STEADY-STATE 150A COLD FILAMENT SURGE RATING.
LENS: MINIMUM 1" DIAMETER, POSITIONED FOR ABOUT 10° VIEW ANGLE.

***Figure 7-20.** Automatic Headlight Dimmer Circuit Using Phototransistor as Light Detector*

7 APPLICATIONS OF PHOTODETECTORS

When increased light falling on the phototransistor causes it to switch to low beams, it is undesirable for only a slight decrease in the level of light to cause a switch back to high beams. If this happened, slight fluctuations in the incident radiation could cause rapid back and forth switching between low and high beams. To prevent this effect, the circuit is designed so that the incident radiation must drop much below the threshold level required for switching to low beams before switching back to high beams. This changing of switching threshold levels is called hysteresis and is often included in a circuit of this type.

Twilight Detector

Junction photodetectors also find applications as twilight detector. In this application, the photodetector is used to detect the surrounding light. When darkness begins to fall, the drop in photocurrent is used to produce some desired effect such as turning on a night light or turning on the parking lights on a car. An example of such a circuit designed to operate from a six-volt battery is shown in *Figure 7-21*. The phototransistor is used to turn the switching transistor on and off. A low light level causes the phototransistor to turn off which allows the base-emitter junction of the switching transistor to be forward biased to turn it on. This effectively connects one side of the lamp to the negative terminal of the six-volt source and the current flowing through the lamp causes it to emit light. A high light level turns on the phototransistor and effectively shorts the base-emitter junction of the switching transistor. This turns off the switching transistor which prevents the flow of current through the lamp.

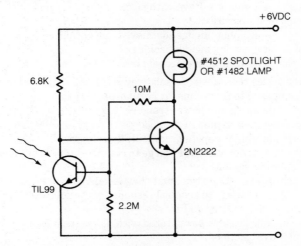

Figure 7-21. Use of Phototransistor in Portable Automatic Nightlight

APPLICATIONS OF PHOTODETECTORS 7

Display Brightness Control

A popular application for photodetectors is to vary the brightness of a television picture to compensate for ambient light changes in the room containing the television receiver. A television scene that is bright enough to be pleasing in a bright room is unpleasantly bright in a dark room. A photodetector placed near the picture tube screen monitors the ambient light and causes a circuit in the television receiver to automatically adjust the display brightness.

Imager

Another possible use for junction photodetectors is in imaging applications. If an array of photodiodes is used, it is possible to focus a portion of an image onto each photodiode in the array. The photocurrent generated by an individual photodiode can be used to indicate the relative brightness of that particular portion of the image. The electrical outputs of the photodiode array can then be easily transmitted to another location for conversion back to an image by a display device. If the image is divided among many photodiodes, good resolution can be obtained. Avalanche photodiodes can even be used to intensify an image detected in dim light.

Energy Conversion

Up to this point, most of our discussion of photodiodes has concentrated on the production of photocurrent. Another possible mode of operation for photodiodes is as a voltage generator, that is, light shining on the photodiode PN junction is used to generate a photovoltage. Using a photodetector to generate a voltage with no external bias is called the photovoltaic mode of operation.

Solar cells are photodiodes that have been optimized for the production of a voltage using incident solar radiation. Some of the ways in which solar cells differ from conventional photodiodes are:

1) the series resistance of a solar cell is typically one ohm or less (compared to fifty ohms for regular photodiodes) in order to provide maximum power transfer to a load,
2) the depletion region in the solar cell is very narrow in order to provide a higher open-circuit output voltage,
3) the sensitive area is very large to compensate for the thinner depletion region,
4) the top layer (the n type material) is kept as thin as possible in order to extend the spectral response of the solar cell into the ultraviolet region, and
5) solar cells are always operated with no external bias (photovoltaic mode).

7 APPLICATIONS OF PHOTODETECTORS

Two of the features of solar cells, large detector area and no biasing requirement, make them very attractive for use in photographic exposure meters and automatic exposure controls.

An important characteristic of solar cells is the solar-cell conversion efficiency. This is the ratio of power out to power in. Present silicon solar cells have a conversion efficiency of ten to twelve percent. The large area required and the resulting high cost limits most commercial applications to watches and small radios at the present time.

SUMMARY OF SELECTION CRITERIA

Let's summarize how one chooses a photodetector for a particular application.

First of all, make sure that the intended application is well defined. Write down as much detail as possible. Classify it so that it can be used to select specifications on photodetectors that are meaningful.

Secondly, what is the source for your optically-coupled system? Are you trying to detect visible light or infrared? Are you trying to detect one wavelength (one frequency) or must you detect a whole spectrum of wavelengths. This will determine the spectral response required of your photodetector so that you can match the photodetector to your source.

Thirdly, how much light will you have at your detector? Will it be by direct radiation from the source or will it come through a transmission medium? Will it be directly from the source you want to detect or will there be unwanted sources around? This will determine the directional sensitivity that you need from a photodetector. It will determine if you need lensed devices. It will determine if you need shields or a special transmission medium. It will prevent interference from sources other than the one you want it to detect.

How about the amount of photocurrent required at the wavelengths? Can you use just a photodiode with its smaller photocurrent output but lower capacitance, high-frequency response and lower dark current, or do you need much higher photocurrent output? If higher output is required, can you use a single phototransistor? It would save space on interconnections and give high reliability at low cost. Compared to the photodiode it has higher dark current, higher capacitance, lower frequency response and the same ratio of I_L to I_D, but still provides a wide linear range of operation and has good design margins for a variety of applications.

If phototransistors do not provide enough photocurrent, then the choice is avalanche photodiodes or PIN photodiodes and external amplifiers, but remember that the avalanche photodiode requires an amplifier with special circuit components and techniques for temperature compensation. The choice also might be influenced by the frequency response required by the system because the amplifier characteristics can be adjusted to obtain the desired frequency response.

APPLICATIONS OF PHOTODETECTORS 7

What is the use of the light? Does it carry data by changing in intensity? If so, is it digital (ON or OFF) or is it analog (continuously varying)? Will it be interrupted and how often? How fast does it have to respond? Answers to questions such as these will determine how fast the system must be and whether it must be analog or digital. If it is digital, most likely the switching times will be of interest. If it is analog, the system frequency response in terms of cutoff frequency will be of interest.

Reliability in terms of consistent long-life performance, low-power, and small size are the advantages offered by solid-state components. These may influence the choice of a photodetector because space or power is limited and environment or long-life determine the choice.

One other point, phototransistors can have the base lead brought out to an external connection. This base lead can be used to turn on or off a phototransistor as if it were an ordinary transistor. Therefore, this additional degree of control for the system is offered by using a phototransistor rather than a photodiode. This could be a significant advantage in a particular application. A phototransistor with the base lead available can also be used as a photodiode. This might mean that the same part can be used for several different applications even though some of the applications require only a photodiode.

WHAT HAVE WE LEARNED?

1) Some photodiode characteristics are electro-optical. Electro-optical characteristics describe the interaction of light energy and its conversion to or control of electrical energy.
2) Some photodiode characteristics are purely optical characteristics. These characteristics describe the way light gets to the sensitive area of the chip.
3) Some photodiode characteristics are primarily electrical in nature and are essentially like those of any semiconductor diode.
4) Phototransistors have essentially the same optical characteristics as photodiodes.
5) Phototransistors differ from photodiodes in that they have higher sensitivity (more gain).
6) Phototransistors with an external base lead can be used in a photodiode mode.
7) The selection of photodetectors for an application is based on required speed of response, level of light available, and amount of photocurrent required.
8) The magnitude of photocurrent required from the photodetector determines if a photodiode, phototransistor, avalanche photodiode, or photodiode with an external amplifier must be used for the application.
9) Applications of photodetectors include many that are obvious, and others that are limited only by the imagination.

7 APPLICATIONS OF PHOTODETECTORS

Quiz for Chapter 7

1. Dark current is
   a. leakage current.
   b. photocurrent.
   c. independent of bias.
   d. independent of temperature.

2. Photocurrent
   a. increases with light intensity.
   b. is larger in phototransistors than photodiodes.
   c. depends on the width of the depletion region (amount of reverse bias).
   d. all of the above.

3. Data sheets for optoelectronic detectors
   a. provide typical data.
   b. differ from manufacturer to manufacturer.
   c. cannot provide data for all possible applications.
   d. all of the above.

4. Phototransistors are used much like any transistor with light acting as
   a. a switch.
   b. additional base current.
   c. additional leakage current.
   d. none of the above.

5. The photodetector for a _____ must have fast response.
   a. light meter.
   b. headlight dimmer.
   c. twilight detector.
   d. photo flash control.

6. Junction capacitance for a photodiode decreases when the diode is
   a. forward biased.
   b. reverse biased.
   c. unbiased.
   d. all of the above.

7. Solar cells have
   a. lower series resistance than photodiodes.
   b. larger area than photodiodes.
   c. no external bias.
   d. all of the above.

8. Phototransistors have
   a. better frequency response than photodiodes.
   b. higher gains than photodiodes.
   c. lower capacitance than photodiodes.
   d. none of the above.

9. Avalanche photodiodes
   a. have higher amplification than photodiodes.
   b. operate in the reverse biased region.
   c. are very temperature sensitive.
   d. all of the above.

10. A device which turns on when exposed to light and stays on until the voltage is reduced is called
    a. phototransistor.
    b. photo field-effect transistor.
    c. photothyristor.
    d. avalanche photodiode.

11. The rise time is _____ proportional to cutoff frequency.
    a. inversely
    b. directly
    c. not
    d. linearly

12. In *Figure 7-4*, the desired current is 30 mA with zero bias. The light would need to be
    a. $50 W/m^2$.
    b. $100 W/m^2$.
    c. $150 W/m^2$.
    d. $200 W/m^2$.

13. In *Figure 7-5a*, the angle between the 50% relative response points is
    a. 15°.
    b. 30°.
    c. 60°.
    d. 90°.

APPLICATIONS OF PHOTODETECTORS 7

14. The rise time and fall times (t_r, t_f) are measured as the time between
 a. 0 and 50% levels.
 b. 10% and 90% levels.
 c. 0 and 100% levels.
 d. 0 and 70% levels.

15. The dark current of a phototransistor is _____ the dark current of the same device used as a photodiode. (See *Figure 7-12*.)
 a. equal to
 b. greater than
 c. less than

16. To use a phototransistor as a photodiode, the two leads used would be the
 a. collector and emitter.
 b. collector and base.
 c. base and emitter.

17. The rise time of a phototransistor is _____ the rise time of the same device used as a photodiode.
 a. greater than
 b. equal to
 c. less than

18. Avalanche photodiodes may have a reference (zener) diode in the same package so that the reference diode can be used to
 a. provide DC voltage.
 b. provide temperature compensation for bias stability.
 c. be used as a current limiter.
 d. be used as a voltage limiter.

19. A photothyristor is a
 a. PNPN device.
 b. photo-sensitive SCR.
 c. fast-switch activated by light.
 d. all of the above.

20. Typical dark currents for photodiodes are in the
 a. ampere range.
 b. milliampere range.
 c. microampere range.
 d. nanoampere range.

1. a, 2. d, 3. d, 4. b, 5. d, 6. b, 7. d, 8. b, 9. d, 10. c, 11. a, 12. b, 13. c, 14. b, 15. b, 16. b, 17. a, 18. b, 19. d, 20. d

8 APPLICATIONS OF PHOTOCOUPLED DATA ACQUISITION SYSTEMS

Applications of Photocoupled Data Acquisition Systems

About This Chapter

In previous chapters, the discussion has emphasized that light performs two services to man. One is the sustenance of life. The other is to bring him information about his environment. This chapter will focus on the applications of photocoupled data acquisition systems. These systems use light sources and light detectors which are designed to obtain and use information about the environment.

WHAT INFORMATION MAY BE OBTAINED ABOUT THE TRANSMISSION MEDIUM?

The transmission medium is the system of matter which forms the path by which light may travel from the light source to the detector. If a source and a detector are appropriately selected and positioned, then systems can be designed to acquire information about position, speed, color, reflectivity, transmissivity, and distances.

Position

A simple example of position detection is shown in *Figure 8-1*. In *Figure 8-1a* the medium consists only of air. The object to be detected does not interfere with the light transmission from the source to the detector, therefore, the current output from the detector is high. In *Figure 8-1b* the opaque object interrupts the light path between the source and detector so the current output from the detector is low. The object becomes a part of the transmission medium so that the light cannot go through. Therefore, an appropriate circuit can utilize the change in detector current to indicate the position of the object, or just the fact that an object is there or not there. The pulses of current generated by ON and OFF light current can be used to count objects as they interrupt the light transmission medium.

Speed

If the object in *Figure 8-1* is continuously moving in the same direction, the speed of its movement can be determined by measuring the time duration of the OFF signal. Since the length of the object is known, the speed is the ratio of the length of the object to the time duration of the OFF signal. For example, if the object length is one inch and the OFF signal duration is one second, the speed is one inch per second.

APPLICATIONS OF PHOTOCOUPLED DATA ACQUISITION SYSTEMS 8

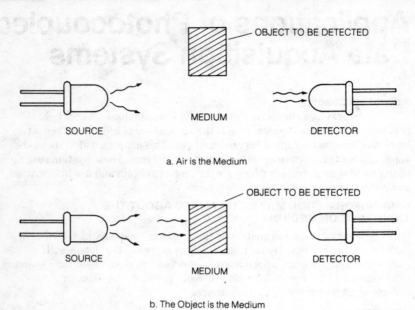

Figure 8-1. Position Sensing

Color

The example shown in *Figure 8-2* shows one way to determine color of the transmission medium. In *Figure 8-2a*, the red light from the source (in this case a red light-emitting diode) shines onto the green filter. The green filter absorbs all wavelengths (color) except green, so none of the red light is transmitted to the detector and the current output from the detector is low. In *Figure 8-2b*, the red light from the source shines onto the red filter which is transparent only to the red wavelengths. Thus, the red light passes through the filter to the detector and produces a high current output from the detector.

Reflectivity

For the purpose of this discussion, *reflectivity* is defined as the ratio of the reflected light to the incident light. *Figure 8-3* shows how reflectivity may be used to acquire information about the medium. In *Figure 8-3a*, 80% of the light incident on the white paper is reflected so a large portion is received at the detector. In *Figure 8-3b*, only 40% of the light incident on the gray paper is reflected so a smaller portion is received at the detector. Therefore, the light reaching the detector in *Figure 8-3a* is about twice that received by the ᵗᵒʳ in *Figure 8-3b*. The result is an electrical signal from the detector ₐₗ to the reflectivity of the paper.

8. Applications of Photocoupled Data Acquisition Systems

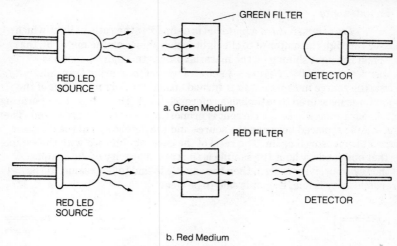

a. Green Medium

b. Red Medium

Figure 8-2. *Color Determination*

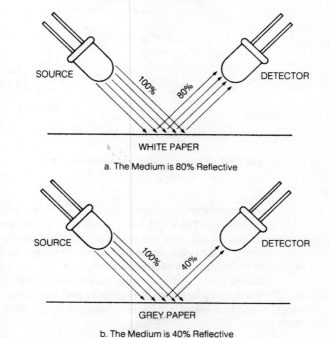

a. The Medium is 80% Reflective

b. The Medium is 40% Reflective

Figure 8-3. *Reflectivity*

APPLICATIONS OF PHOTOCOUPLED DATA ACQUISITION SYSTEMS

Transmissivity

Transmissivity of a material is defined as the ratio of light which passes through the material to the light which shines on the material; that is, if half the light shining on the material passes through, then it has a transmissivity of 50%. *Figure 8-4* shows a way of measuring transmissivity. First, the source in *Figure 8-4a* is turned on and the current output of the detector is measured to establish a reference level. The detector is operating in a linear range where the current is proportional to the light received. Then the sample is placed between the source and the detector, and the detector current is measured again. The ratio of the current obtained with the sample to that obtained without the sample is the transmissivity of the sample. If the transmissivity is zero, then the sample is said to be opaque. If the transmissivity is one, then it is said to be clear.

a. Reference Level is Established with Only Air as Medium

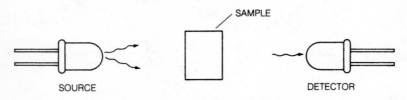

b. Sample Inserted as Medium

***Figure 8-4.** Transmissivity Measurement*

Distance

If a flat lensed source with a constant output and a wide viewing angle is used to illuminate a detector which has a specified sensitive area, a relationship can be established between distance d (*Figure 8-5*) and the light intensity at the surface of the detector. Since the current through the detector increases as the intensity increases, a relationship between distance and current can be established. If the light source emits light uniformly in all directions, the relationship between the intensity and distance is approximated by the equation below, where E_e is the intensity in W/m^2 at distance d in inches.

$$K = E_e \times d^2 \qquad (1)$$

8 APPLICATIONS OF PHOTOCOUPLED DATA ACQUISITION SYSTEMS

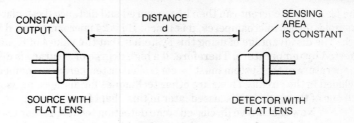

Figure 8-5. Distance Measurement

If d is set at one inch and the current through the source is adjusted to obtain an intensity (E_e) of $1W/m^2$, then K may be calculated as shown below by substituting the values in equation *1*.

$$K = 1 \times 1^2$$
$$K = 1$$

By substituting K = 1 in equation *1*, the relationship between d and E_e may be expressed as:

$$E_e \times d^2 = 1$$

or

$$d = \sqrt{\frac{1}{E_e}} \qquad (2)$$

Since most photodiodes have a linear response to intensity, the intensity can be related to the photodiode current as shown in equation *3*:

$$E_e = M \times i \qquad (3)$$

where i is the photodiode current in mA and M is the slope of the irradiance versus reverse current curve. Substituting $M \times i$ for E_e in equation *2*, the following relationship is obtained:

$$d = \sqrt{\frac{1}{M \times i}} \qquad (4)$$

M is the ratio of irradiance in W/m^2 to photocurrent (reverse current) in amperes. The value for M can be obtained from the curve of photocurrent (reverse current) versus irradiance for a particular detector given on a data sheet. For example, refer to *Figure 7-3*. Note that when the irradiance is $1W/m^2$, the reverse current is approximately 6 microamperes. Therefore, the value of M for this detector is $1W/m^2$ divided by 6 microamperes, or $M = 1.66 \times 10^5$. Substituting this value of M in equation *4* gives

$$d = \sqrt{\frac{1}{1.66 \times 10^5 \times i}} \qquad (5)$$

APPLICATIONS OF PHOTOCOUPLED DATA ACQUISITION SYSTEMS

8

The photodiode current can then be measured and distance d calculated. For example, if i is 6 microamperes, d is 1 inch. If i is 24 microamperes, d is 0.5 inch. The disadvantage of using this system is that the values of E_e and M are affected by temperature. Therefore, if a high degree of accuracy is desired, the temperature of the system must be controlled or temperature compensation included in the circuit. There are other techniques for distance measurement and one of them will be discussed later in this chapter.

Now that we've discussed some detection techniques for transmissive systems and reflective systems, let's look further at some typical applications of each.

WHAT ARE SOME TYPICAL APPLICATIONS OF TRANSMISSIVE SYSTEMS?

Punched Card Readers

Punched card readers were among the first high volume applications of semiconductor light sources and detectors. Punched data cards (often called IBM cards), such as the one shown in *Figure 8-6*, were (and still are) used to store computer programs or data for input to the computer. The data on the card is represented by holes punched in the 80 columns by 12 rows. Therefore each column can contain 12 bits of information.

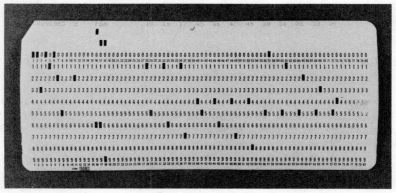

Figure 8-6. IBM Data Card

The first technique used to read the card was to run the card through a reader consisting of 12 mechanical switches that made contact through the holes as the card passed through the reader. The contacts detected the location of the holes in the column and this information was passed to the computer. A major problem with this approach was poor reading reliability caused by dust and paper shavings interfering with the switch contacts. Optical readers, such as those shown in *Figure 8-7*, are typical of those that were developed to read these cards with much higher reliability. The TIL136 is a pair of circuit boards, one containing 12 TIL23 infrared LEDs as sources and the other containing 12 TIL602 phototransistors as detectors.

8. APPLICATIONS OF PHOTOCOUPLED DATA ACQUISITION SYSTEMS

TIL134 . . . 12-ELEMENT GALLIUM ARSENIDE IRED ARRAY
TIL135 . . . 12-ELEMENT PHOTOTRANSISTOR ARRAY
TIL136 . . . 12-CHANNEL PAIR

- Center-to-Center Spacing of 6,3 mm (0.250 Inch) for Tape Reading
- Reliable Solid-State Components
- IRED's Eliminate Lamp-Filament-Sag Problems
- Spectrally Matched for Improved Performance
- Printed Circuit Board Construction Allows Precise Alignment

description

The TIL134 is an array of twelve TIL23 gallium arsenide infrared-emitting diodes mounted in a printed circuit board. The TIL135 is an array of twelve selected LS600 phototransistors. The TIL136 is a pair of selected arrays comprising a TIL134 and TIL135 and offering guaranteed channel performance.

mechanical data

The printed circuit board material is glass-base NEMA standard FR-4, class II, 0,6-kg/m^2 (2-oz/ft^2) copper clad on each side. The approximate weight of the TIL134 and TIL135 is 8.5 grams each.

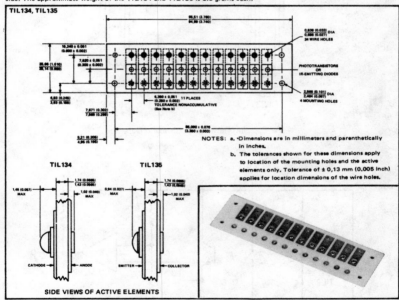

Figure 8-7. Punched Card Reading Devices

APPLICATIONS OF PHOTOCOUPLED DATA ACQUISITION SYSTEMS   8

The LED and detector are positioned with a 0.200 inch spacing between the lenses of the devices as shown in *Figure 8-8a* to provide space for the card to pass between the pair. When a punched hole is positioned over the detector, the current through the detector is guaranteed to be above 2.5mA but less than 10mA when the LED source is driven at 50mA. A circuit shown in *Figure 8-8b* may be used to supply a high-level output when light is passed through a punched hole or a low output when light is blocked by the card. Of course, twelve of these circuits are required to read a card.

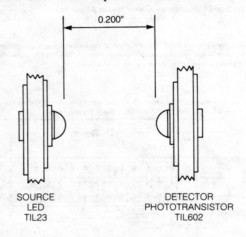

a. Spacing Between Source and Detector

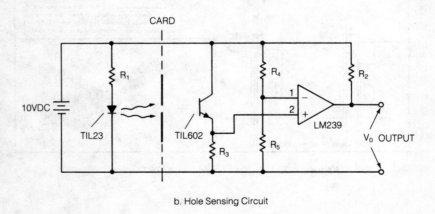

b. Hole Sensing Circuit

Figure 8-8. *Single Channel Punched Card Reader*

8 APPLICATIONS OF PHOTOCOUPLED DATA ACQUISITION SYSTEMS

The LM239 is a differential comparator circuit like the one discussed in Chapter 6. When light strikes the TIL602 detector the voltage across R3 increases due to increased photocurrent. This is the voltage that is applied to input 2 of the comparator. When the voltage at input 2 is greater than the voltage at input 1, formed by the voltage divider of R4 and R5, the output of the comparator is driven to a high-level voltage. Otherwise, with no light on the detector the input 1 voltage is greater than the input 2 voltage and the output voltage level of the comparator is low.

Position Sensing

Position sensing is perhaps the most common use of optically coupled systems. A simple example was given earlier in *Figure 8-1*. This technique can be used to detect the presence of objects between the source and the detector. Position detectors of various types are used in copiers, printers, computer disk drives, keyboards, shaft encoders, and many other applications.

Also, as mentioned earlier, this detector can detect speed as well as presence. *Figure 8-9* shows a position detector that lights a warning indicator when a shaft slows below its normal operating speed. As shown in *Figure 8-9a*, the system is made up of a motor driving an encoder disk between an assembly that contains both a source and a detector. When the open slit in the encoder passes between the detector and the source of the TIL143, the increase in detector current causes an increase in voltage at the inverting input of comparator 1. R_2 and R_3 are selected to assure that the inverting input voltage level swings past the reference voltage level established at the noninverting input. The output of comparator 1 then goes to a low level which discharges capacitor C_1. As the encoder continues to move, the slit passes and the light from the source is blocked from the detector. Detector current decreases and the inverting input level to comparator 1 decreases below the noninverting input. The output of comparator 1 goes to a high voltage level and the capacitor begins to charge at a rate determined by the RC time constant of R_4 and C_1. If another slit in the encoder causes another detector current pulse before capacitor C_1 charges to the reference voltage determined by R_5 and R_6, the output of comparator 2 stays at the high voltage level so the VLED will remain off. If, however, the shaft slows down sufficiently, C_1 will have time to charge between pulses to a voltage level sufficient to cause the output voltage of comparator 2 to switch to a low level. The VLED will be turned on to indicate the slow shaft speed. Modifications could be added to the system to turn off the motor, or, in addition to turning on the warning VLED, some other control circuit could be turned on to cause something else to happen.

APPLICATIONS OF PHOTOCOUPLED DATA ACQUISITION SYSTEMS 8

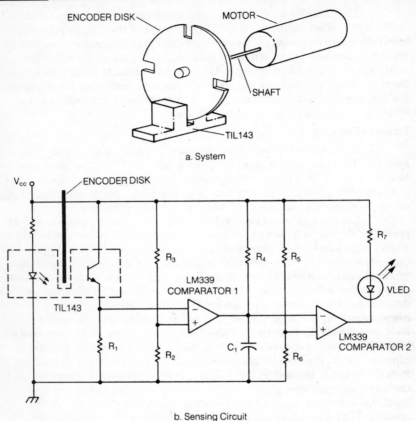

Figure 8-9. Speed Sensing

Fluid Level Sensing

The level of fluid in a container can be sensed with a source-detector pair as shown in *Figure 8-10*. As the fluid level reaches the level of the source-detector, the transmission medium (light path) changes from air to the fluid. If the transmissivity of the fluid is different from air, the amount of light received by the detector will be reduced. If the fluid does not block enough light to reduce detector current sufficiently, a float could be used to improve response. The detector output can be used to control the fluid input to prevent an overflow and/or turn on a visual indicator lamp to show that the container is full. Source-detector pairs and indicator lamps could be used at different levels, if desired, to indicate that the fluid is at levels other than full.

8. APPLICATIONS OF PHOTOCOUPLED DATA ACQUISITION SYSTEMS

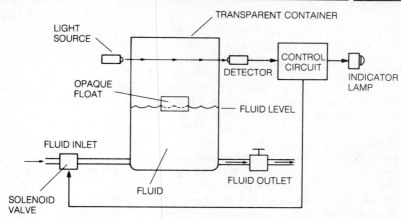

Figure 8-10. *Fluid Level Detector Using Fluid to Block the Light Path*

Another technique is shown in *Figure 8-11*. The source and detector are mounted in a rod that is transparent to the wavelength of the source light and the rod has a wedge on the bottom formed by a 90 degree angle. If the fluid level is below the wedge, the light from the source is reflected back to the detector. When the fluid level reaches the wedge, the index of refraction changes and the light continues traveling straight into the fluid. (The index of refraction will be discussed in Chapter 9.) Only a small portion of the light is reflected back to the detector. This approach differs from that in *Figure 8-10* because the fluid *changes the transmissive path* rather than changing the *transmissive characteristics of the path*.

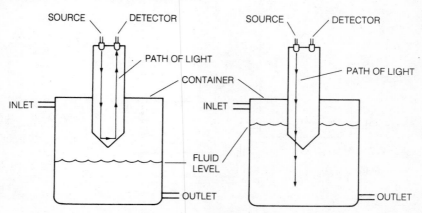

Figure 8-11. *Fluid Level Detector Using Fluid to Change the Light Path*

APPLICATIONS OF PHOTOCOUPLED DATA ACQUISITION SYSTEMS

8

Figure 8-12 shows a typical circuit which could be used to maintain fluid between two levels. When the fluid drops below level B, detector S_2 provides a current that produces a positive going voltage pulse to the non inverting input of comparator 2. Comparator 2 produces a logic high-level voltage at the set input of the RS flip-flop which sets the flip-flop output Q to a logic high-level voltage. When Q goes to the logic high-level voltage, the buffer amplifier passes a control signal to turn on the pump and the fluid level begins to rise. As soon as the level goes above B, detector S_2 is turned off, the comparator switches back and the set input to the flip-flop goes low. However, Q remains at a logic high-level voltage level because the reset input is also at a logic low-level voltage. When the fluid level reaches A, detector S_1, which has been producing photocurrent, turns off and provides a negative going voltage pulse at the inverting input of comparator 1. As a result, the output voltage level of comparator 1 goes to a logic high-level voltage. This resets the RS flip-flop and the voltage level of Q goes to a logic low-level voltage. When Q goes low, the control signal turns off the pump. When the fluid level drops below A, detector S_1 again produces photocurrent to raise the voltage on the inverting input of comparator 1. Comparator 1 switches and the input voltage to the reset input of the flip-flop goes to a logic low-level voltage, but Q remains at the logic low-level voltage. Therefore, the pump remains off until the fluid level drops below level B again.

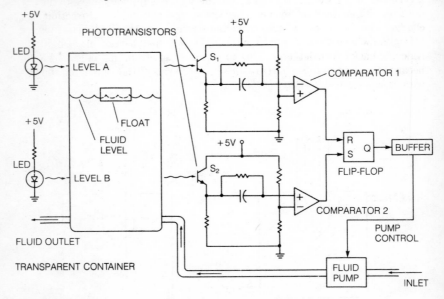

Figure 8-12. *Typical Circuit for Controlling Fluid Between Two Levels*

8 APPLICATIONS OF PHOTOCOUPLED DATA ACQUISITION SYSTEMS

Brushless Motors

A simplified diagram of a dc motor is given in *Figure 8-13*. Current from the dc power source flows through the carbon brushes to the copper commutator bars. The brushes are stationary but the commutator bars rotate with the armature and shaft. From one commutator bar, the current flows into a coil of wire wound around the armature and back out the other commutator bar and brush to the dc power source. This current sets up a magnetic field which is perpendicular to the magnetic field set up by the permanent magnet. As the two magnetic fields try to align, a torque is set up which causes the armature to rotate. As soon as the armature rotates enough for the magnetic fields to align, the commutator bars switch positions beneath the brushes. This reverses the direction of the current through the armature which causes the direction of the magnetic field produced by the armature to reverse. The two magnetic fields are again out of alignment and as they try to align, the rotation is continued in the same direction.

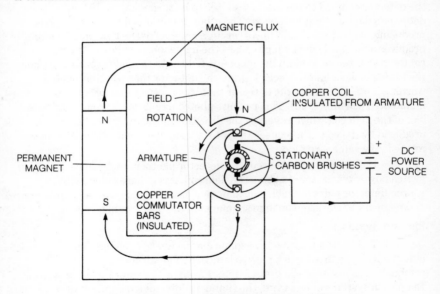

Figure 8-13. *Simplified Diagram of DC Motor*

APPLICATIONS OF PHOTOCOUPLED DATA ACQUISITION SYSTEMS 8

Although a dc motor such as this would work, it is not self-starting since there is a "dead spot" where no torque is produced at the time the brushes are switching between commutator bars. Practical dc motors use several coils with a pair of commutator bars for each coil. With this arrangement, at least one of the coils is producing torque while others are switching.

For some applications in explosive gas environments, the electrical arcs generated by the current switching between the stationary brushes and rotating commutator are not acceptable. The arcing also damages both the commutators and brushes so that regular maintenance is required.

Optoelectronic devices may be used to switch the currents without requiring brushes or other physical contact. A simple example is shown in *Figure 8-14*. In this example, the transmission path for the light from one of the LED sources is blocked by a timing disk that has one-half (180 degree segment) clear and the other half (180 degree segment) opaque. This causes the detector coupled to the blocked source to be turned off so that only the detector's dark current is flowing. The other detector is illuminated so it is producing photocurrent that flows through one-half of the field coil. The coil produces a magnetic field which causes the moveable permanent magnet to rotate in an attempt to align the magnetic fields. As the fields reach alignment, the timing disk suddenly blocks light to the detector that was illuminated (turning it off) and allows the detector that was receiving no light to be illuminated (turning it on). This permits photocurrent flow through the other half of the field coil which reverses the direction of the magnetic fields produced by the coil. This causes the rotor (permanent magnet) to continue to rotate in the same direction.

If more power is desired, the detector photocurrent can be used to switch power transistors to drive the field coil. Also, more poles, with two source-detector pairs for each pole and appropriate changes in the number and arrangement of clear and opaque segments in the disk, could be added.

Ignition Systems

The basic function of a gasoline engine ignition system for automobiles is to fire a spark plug to ignite the fuel-air mixture in the combustion chamber at the proper time. *Figure 8-15* shows the basic system. In ignition systems of this type, the timing is determined by mechanical linkages. The last link is the lobed distributor shaft pushing on an insulated rubbing block attached to the moveable contact of the switch contact set. These contacts are usually referred to as the "points". When the points are closed, a heavy current flows in the primary of the ignition coil. When the points open, the collapsing field in the primary produces a high voltage of around 15,000 volts in the secondary which causes a spark at the tip of the

8 APPLICATIONS OF PHOTOCOUPLED DATA ACQUISITION SYSTEMS

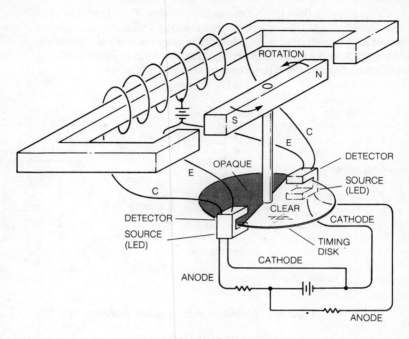

a. Physical Arrangement

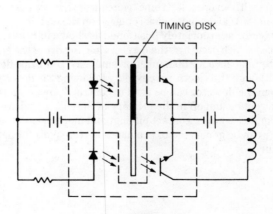

b. Schematic Diagram

Figure 8-14. Brushless DC Motor

APPLICATIONS OF PHOTOCOUPLED DATA ACQUISITION SYSTEMS

8

spark plug to ignite the fuel mixture in the cylinder. The disadvantages of this system are pitting of the contact points caused by arcing, contact bounce of the points at high speeds, change in timing caused by mechanical wear of the rubbing block, and a lower high voltage than optimum because of current limitation of the points and the mechanical inertia of the system.

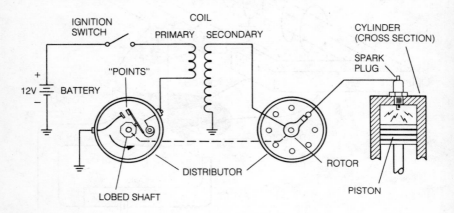

Figure 8-15. *Automobile Ignition System*

Some of these disadvantages can be overcome by using an optically coupled system with an optoelectronic transducer that senses position to replace the points. If an encoder disk is placed on the distributor shaft with a source and detector appropriately positioned as shown in *Figure 8-16*, a signal can be produced which occurs at the correct time for firing the spark plug. The source-detector and encoder disk function similarly to the one described in *Figure 8-9*, although the electronic circuitry following the detector is different. The signal from the detector can be used to control a power transistor to switch the current on and off in the primary of the coil. Besides eliminating the problems of contact bounce, contact pitting, and mechanical wear; a much higher voltage can be generated to produce a better spark for better ignition of the fuel mixture.

8. APPLICATIONS OF PHOTOCOUPLED DATA ACQUISITION SYSTEMS

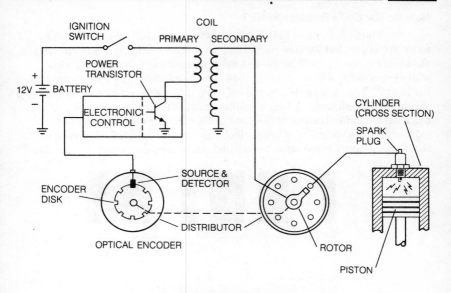

Figure 8-16. Optical Ignition Encoder

WHAT ARE SOME TYPICAL APPLICATIONS OF REFLECTIVE SYSTEMS?
Bar Code Readers

Bar code readers convert a printed bar pattern into digital signals for computers. They are used for inventory control systems in retail stores and in industry. They also are used to load programs into computers and programmable calculators. The computer programs are coded and printed on paper which may be easily distributed to many users. The program is loaded into the computer or calculator by passing the bar code reader (BCR) across the code.

Almost all products that are widely distributed for retail sale are identified by a bar code called the universal product code. Many supermarkets and department stores have a central computer programmed to translate the code to the product name and assign the price. A bar code reader is used at each checkout counter to read the code and feed it to the computer. The computer then causes the item name and current price to be displayed on the terminal and printed on the customer's receipt. Also, the quantity of the item purchased is subtracted from the inventory record in the computer for an instant update. When the inventory for that item falls below a predetermined level, the computer can alert the manager through the inventory report, or will automatically print out an order for more items if that function is desired.

APPLICATIONS OF PHOTOCOUPLED
DATA ACQUISITION SYSTEMS

8

How Do Bar Code Readers Work?

The first element in the system is the code itself. Many different codes are in use, but for this discussion, the simple black and white bar code shown in *Figure 8-17* will be used. A narrow black bar followed by a wide white bar is defined to be a "1". A wide black bar followed by a narrow white bar is a "0". The "1" and "0" are the only two states allowed by the basic binary digital signal definition. A long black bar equal to the width of two regular black bars signifies the end of the code. An optical bar code reader system must contain an optical read head which has a small enough field of view so that it can detect the narrowest black and white lines contained in the code.

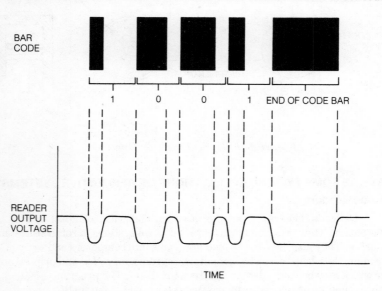

Figure 8-17. Bar Code and Bar Code Reader Output

Among the first users of bar codes were the railroad companies. The codes are painted on the sides of the cars to identify the car and aid in delivery to its destination. These bars are large and can be read from several feet away as the cars pass by the reader. However, the codes found on the small packages in retail stores may require accurate detection of bars only 0.010 inch wide.

8. APPLICATIONS OF PHOTOCOUPLED DATA ACQUISITION SYSTEMS

Figure 8-18 demonstrates a hand-held bar code reader (called a wand) being used to read a bar code. Light from the source reflects from the mirror to the surface of the bar code. The detector is behind an aperture and a field stop which limits the field of view of the detector so that the area of the surface "seen" by the detector is narrower than the smallest bar space to be detected. This is illustrated in *Figure 8-19*.

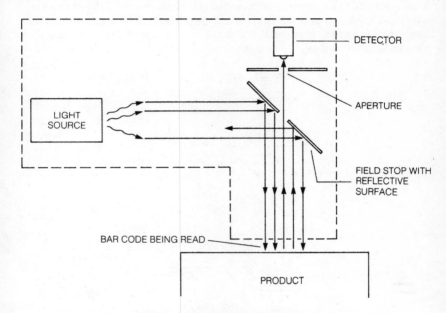

Figure 8-18. Handheld Bar Code Reader

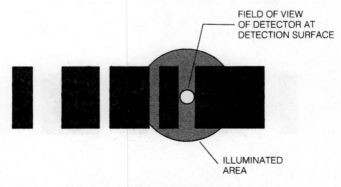

Figure 8-19. Detector's Field of View

APPLICATIONS OF PHOTOCOUPLED DATA ACQUISITION SYSTEMS 8

Figure 8-20 demonstrates a different technique often used in systems for supermarkets. This reader is not handheld but is placed below a window in the counter surface. Light from the laser reflects from the mirror and is swept across the bar code by the rotation of the mirror. The requirement of this system is that the beam from the laser must illuminate a spot on the surface which is smaller than the narrowest bar to be read as shown in *Figure 8-21*. This is not difficult to achieve because the laser beam can be focused to a very small spot. Notice that in the first case (*Figure 8-19*) the illuminated area was large and the sensing area small, but in the second case (*Figure 8-21*) the sensing area is large and the illuminated area small.

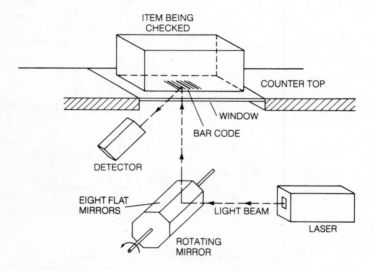

Figure 8-20. Laser Scanning Bar Code Reader

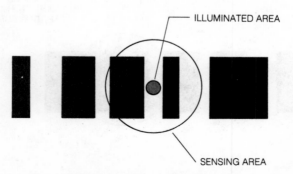

Figure 8-21. Area Illuminated by Laser

8 APPLICATIONS OF PHOTOCOUPLED DATA ACQUISITION SYSTEMS

The two techniques described each have advantages and disadvantages. The laser scanning technique allows the system to scan the code over and over as long as the item is held in place. This yields better accuracy and reliability in reading, particularly when the bars are smeared or damaged. With the handheld wand, repeated scans are possible only if the operator performs it. However, handheld wands are relatively low cost when compared to the laser scanning system. The laser systems do not require physical contact to the surface containing the code while most handheld scanners do require contact.

When the opto-electronic problem is solved; that is, an appropriate read head is available which produces a signal such as the one shown in *Figure 8-17*, then the remainder of the system can be selected or designed. *Figure 8-22a* shows a functional block diagram of the system.

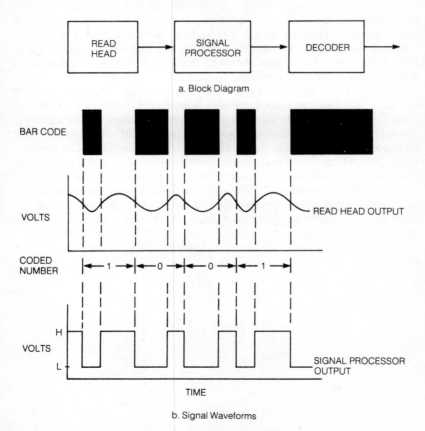

Figure 8-22. *Bar Code Reader System*

APPLICATIONS OF PHOTOCOUPLED DATA ACQUISITION SYSTEMS

8

The read head transforms the bar code into electrical signals that vary in voltage level from high to low. The signal processor reshapes and squares up the read head output signal to ensure reliable operation of the following circuits. The processor also shifts the signal level so the processor output signal has the required digital levels. The signal waveforms are shown in *Figure 8-22b*. The decoder for the code shown would compare the length of the black pulse to the length of the white pulse and decode the signal based upon the rules of the code selected. As previously stated, the rules in this example are that a long black pulse followed by a short white pulse is a "0", and a short black pulse followed by a long white pulse is a "1". The code detected in *Figure 8-22* is 1001.

Range Finding Systems

The simple range finding system discussed earlier in *Figure 8-5* uses the variations of light intensity with distance. A better system uses the phase shift of a reflected infrared signal. Although this method is limited to short ranges (1 to 15 meters), it is useful for applications such as automatic focusing controls for cameras, fluid level detectors that do not require contact with the fluid, and proximity alarm systems.

The basic concept of a phase-shift range finding system is illustrated in *Figure 8-23*. The modulator provides a frequency (f_m) which modulates (varies) the intensity of the infrared radiation from a source such as a gallium arsenide LED. The modulator also sends the same signal to a phase comparator which sets a flip-flop at a zero crossing of the modulating signal. The radiation from the IR transmitter travels to the object and is reflected to the IR receiver. A high-speed silicon photodiode responds to the reflected light which is slightly delayed with respect to the signal from the modulator to the phase comparator because of the distance to the object and back to the receiver. The photodiode output signal resets the flip-flop in the phase comparator so that the output is a pulse whose width is proportional to the delay time. The time delay, t_d, shown in *Figure 8-23b*, is expressed in equation 6.

$$t_d = \frac{2d}{(3 \times 10^8)} \qquad (6)$$

8. APPLICATIONS OF PHOTOCOUPLED DATA ACQUISITION SYSTEMS

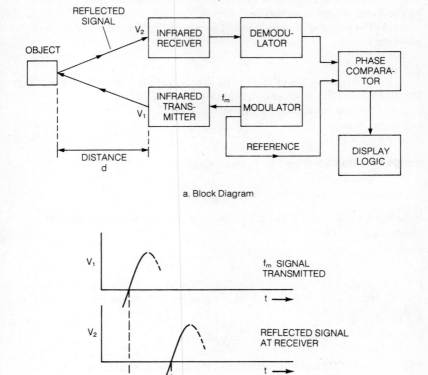

a. Block Diagram

b. Signal Waveforms

Figure 8-23. Phase Shift Range Finder System

APPLICATIONS OF PHOTOCOUPLED
DATA ACQUISITION SYSTEMS

The time delay in seconds is twice the distance to the object in meters divided by the speed of light in meters per second.

The phase shift in radians of the modulating frequency is given in equation 7.

$$\phi = 2\pi f_m t_d \qquad (7)$$

It is equal to the number of radians in 360°(2π) times the modulating frequency times the time delay t_d that is measured.

Solving equation 6 for d gives equation 8.

$$d = \frac{(3 \times 10^8) t_d}{2} \qquad (8)$$

Solving equation 7 for t_d gives equation 9.

$$t_d = \frac{\phi}{2\pi f_m} \qquad (9)$$

Substituting equation 9 for t_d in equation 8 gives the range in terms of the phase angle as follows:

$$d = \frac{(3 \times 10^8)}{2} \times \frac{\phi}{2\pi f_m}$$

Since there are 360 degrees in 2π radians then 360° can be substituted for 2π and the distance d in meters is given in equation 10 in terms of the phase angle in degrees that is detected.

$$d = \frac{3 \times 10^8}{720 f_m} \phi \qquad (10)$$

Since the phase comparator is limited to the 180 degrees between zero crossings of the signal, the range is limited by the frequency of the modulating signal. For example, if the modulation frequency is 5MHz, the maximum range is 15 meters.

$$d = \frac{(3 \times 10^8)(180)}{720 \times 5 \times 10^6}$$
$$= \frac{3 \times 10^8}{20 \times 10^6}$$
$$= 15 \text{ meters}$$

The same techniques can be used for moving targets in the radiated field.

8. Applications of Photocoupled Data Acquisition Systems

Guidance Systems

Guidance systems of many kinds have been designed using optical sources and detectors. Many military systems use light, particularly infrared, to guide missles or bombs to their targets. To develop a concept of how these systems work, consider *Figure 8-24* which shows two toy army tanks. They are designed so that one tank follows and chases the other tank. The chase tank has a pair of detectors which can respond to the light on the enemy tank.

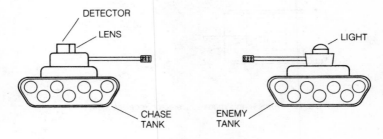

Figure 8-24. Toy Tanks

When the light (target) is directly in front of the chase tank, the light is focused equally on the two photo diodes as shown in the top view in *Figure 8-25a*. When the light is to the right of the tank, the light is focused on the left detector as shown in *Figure 8-25b*. When the light is to the left of the tank, the light is focused on the right detector as in *Figure 8-25c*. Two motors are used in the chase tank; one to power the right track, the other to power the left track. The right motor speed is increased and the left decreased when more light is shining on the right detector and, conversely, the left motor speed is increased and the right decreased when the light is brighter on the left detector. Both motors are controlled to the same speed if the light level is the same on both detectors. Once the light (target) is in view of the chase tank detectors, the detector control of the motors will constantly correct the direction of travel to follow the light on the enemy tank.

Much more complicated systems, built with elaborate feedback systems use these basic concepts to guide missiles, bombs, and aircraft, to aim and fire guns, and to direct and control vehicles. The use of light, particularly infrared, has indeed resulted in many useful and practical applications.

APPLICATIONS OF PHOTOCOUPLED DATA ACQUISITION SYSTEMS 8

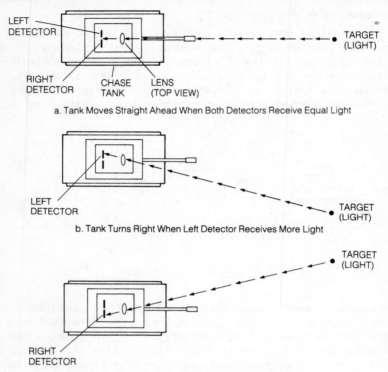

Figure 8-25. Detectors Control Tank Movement

WHAT HAVE WE LEARNED?

1) Optoelectronics can be used to determine characteristics of the transmission medium such as position, speed, color, reflectivity, transmissivity, and distance.
2) The transmission medium can be controlled to contain retrievable data (data that can be recovered) by use of optoelectronics systems to detect characteristics of the transmission medium. Examples are computer data cards, paper tape, and reflective codes such as bar codes.
3) Optoelectronics systems can be used to sense fluid levels or replace brushes in electric motors.
4) Optoelectronics can be used to measure distance or direction.
5) Optoelectronic systems can be developed to control machines by adding a measure of vision (sensory perception) to the machine to allow the machine to be controlled or guided in its actions based on inputs to its detectors from the surrounding environment.

8 APPLICATIONS OF PHOTOCOUPLED DATA ACQUISITION SYSTEMS

Quiz for Chapter 8

1. The function of photocoupled data acquisition systems is to
   a. save energy.
   b. detect sound waves.
   c. acquire information about the light source.
   d. acquire information about the transmission media.

2. The purpose of a punched card reader is to
   a. detect the position of a card.
   b. retrieve data stored on the card.
   c. detect the color of the card.
   d. detect the velocity of the card.

3. The advantage of a brushless motor is
   a. reliability.
   b. low service requirements.
   c. no spark generation.
   d. all of the above.

4. The advantage of optoelectronic components in the ignition system is
   a. no point bounce.
   b. no mechanical wear.
   c. no stress on the distributor shaft.
   d. all of the above.

5. The advantage of optical card readers is
   a. speed.
   b. no mechanical contact of detector to card.
   c. less sensitivity to dust and paper particles.
   d. all of the above.

6. The advantage of optoelectronic devices in fluid level sensing is
   a. no moving parts.
   b. no mechanical contact with fluid.
   c. fast response to fluid level changes.
   d. all of the above.

7. An optical bar code reader must
   a. detect speed.
   b. have strong ambient light.
   c. detect differences in reflectivity of the printed bars and spaces.
   d. none of the above.

8. The transmissivity of a certain material is 100% for green light. The material is
   a. opaque.
   b. transparent at all wavelengths.
   c. transparent to green light.
   d. reflects green light.

9. For an intrusion (burglar) alarm, the best light source for a photocoupled system would be a
   a. incandescent lamp.
   b. VLED.
   c. infrared LED.
   d. neon lamp.

10. The current produced by a photodiode is
    a. directly proportional to distance from the light.
    b. inversely proportional to distance from the light.
    c. proportional to the square of the distance from the light.
    d. inversely proportional to the square of the distance from the light.

11. A red optical filter _____ red light.
    a. reflects
    b. absorbs
    c. passes
    d. amplifies

12. A red optical filter is used with one detector and a green optical filter is used with another detector. The outputs of the detectors are the same. The color of the source contains _____ light.
    a. only red
    b. only green
    c. approximately equal amounts of red and green

13. A bar code must contain
    a. at least two colors.
    b. at least three colors.
    c. only two colors.
    d. all of the above.

APPLICATIONS OF PHOTOCOUPLED DATA ACQUISITION SYSTEMS

8

14. The amount of data in a bar code is limited by
    a. the light intensity.
    b. the field of view of the detector.
    c. the color of the light.
    d. the color of the code.

15. In the fluid level detector shown in *Figure 8-10*, the float is used
    a. to block the light.
    b. must be opaque.
    c. is not necessary if the fluid is opaque.
    d. all of the above.

16. In the card reader circuit shown in *Figure 8-8*, the comparator is used to
    a. prevent ambient light from producing an output.
    b. provide the proper output voltage levels.
    c. provide a "clean" output pulse shape.
    d. all of the above.

17. The laser bar code reader illustrated in *Figure 8-20*
    a. is faster than the hand held wand.
    b. can read smaller codes than the wand.
    c. costs more than the wand.
    d. all of the above.

18. The purpose of the two detectors in the toy tank of *Figure 8-25* is to
    a. determine the relative direction of the light.
    b. determine the intensity of the light.
    c. determine the distance to the light.

19. In the fluid level detector of *Figure 8-11*, the wedge must be 90° so that
    a. the light is reflected back to the detector.
    b. the light can be transmitted through the fluid.
    c. the fluid will not cause reflection to the detector.

20. The range of the optical range finding system of *Figure 8-23* is determined by
    a. the frequency (wavelength of the light).
    b. the frequency of the modulating signal.
    c. the reflectivity of the object.

1. d, 2. b, 3. d, 4. d, 5. d, 6. d, 7. c, 8. c, 9. c, 10. d, 11. c, 12. c, 13. c, 14. b, 15. d, 16. d, 17. d, 18. a, 19. a, 20. b

9 APPLICATIONS OF PHOTOCOUPLED DATA TRANSMISSION SYSTEMS

Applications of Photocoupled Data Transmission Systems

ABOUT THIS CHAPTER

This chapter is about photocoupled data transmission systems. Photocoupled data *transmission* systems are those systems which pass information from the source to the detector. These systems, in contrast to the photocoupled data *acquisition* systems, are not designed to determine any characteristics of the transmission medium (environment) but rather to receive information transmitted to the detector by the source. The medium or environment is used simply as a path by which the light signals may travel from the source to the detector. Care must be taken, therefore, to evaluate the medium through which the light carrier and its information signals pass to ensure that the transmitted signals are not blocked, attenuated, or masked by unwanted signals (noise).

WHAT TYPES OF TRANSMISSION MEDIUM ARE USED?

Virtually any medium through which light can travel can be used. Three types of medium will be discussed: air, transparent materials, and fiber optics.

Air

Air is often the total transmission medium or at least a large part of the transmission path. At first glance, air seems to be an excellent medium through which light can travel. However, when the transmission distance is greater than half a mile, two problems may become evident. One problem is that air attenuates light and the attenuation is even greater in fog, rain, snow, and dust. As has been discussed earlier, sunlight is attenuated by the earth's atmosphere from *1,400 watts/m²* in space to *800 watts/m²* at the earth's surface.

The other problem is the varying temperature of air causes the density of the air to vary. This can cause the light to change directions so that the light path is altered. This is the effect which causes a mirage, and it has been observed by almost everyone. Remember as you drive down a highway, especially on a hot day, a mirage makes the road ahead of you appear as if it were covered with water. This is because of the altered light paths. If a laser were used to send a signal over a distance of one mile as shown in *Figure 9-1*, the light path from the laser could be altered enough by the varying densities of the air to miss the receiver entirely. Also, a problem for long distance light transmission is that the transmission path is limited to the line-of-sight path.

APPLICATIONS OF PHOTOCOUPLED DATA TRANSMISSION SYSTEMS

9

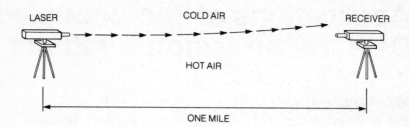

Figure 9-1. Bending of Light Because of Air Temperature

Because of these problems, a medium over which more control may be exercised is needed. The medium often chosen for light transmission is fiber optics. To understand the use of fiber optics we need to understand more about how light propagates through transparent materials.

Transparent Materials

Any material which is transparent to the wavelength of the light being used is a possible medium for a data transmission system. Remember that the transmission medium does not have to be visually transparent if, for example, infrared wavelengths are used.

A photocoupler (opto-isolator), which contains both a light source and a light detector in the same package, may use plastic as the transmission medium as shown in *Figure 9-2*. Light is generated by the LED and tends to travel through the clear plastic overcoat to the detector. In this example, it can be seen that the light which travels directly to the detector does not strike the detector on its light sensitive top surface. Other light rays travel to the outer surface of the clear plastic and would tend to be absorbed by the black plastic encapsulation that forms the outer package. To prevent this, the medium is altered by coating the clear plastic with a white reflective film. The result is that the light propagating to the outside surface of the clear plastic is reflected and a significant improvement is noted in the amount of light reaching the active portion of the detector.

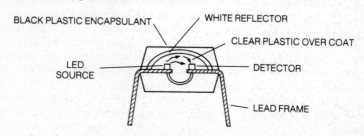

Figure 9-2. Typical Photocoupler Construction

9 APPLICATIONS OF PHOTOCOUPLED DATA TRANSMISSION SYSTEMS

The important observation is the transmission medium can be controlled to improve the efficiency of the system. In addition to the actual material selection, the transparent material can be shaped to achieve gains in efficiency. These transmission paths are called light guides and can be formed to deliver light from the source to the detector through paths that are not straight or aligned. One very important light guide for optically coupled systems is referred to by the term fiber optics.

Fiber Optics

Angle of Incidence and Reflection

In order to discuss fiber optics light guides, two important facts about optics must be introduced. In *Figure 9-3*, notice that one line is at a right angle (perpendicular) to a smooth reflective surface. This line is called the *normal*. When a light ray strikes the surface from any direction other than perpendicular (straight in), it forms an angle with respect to the normal. This angle is called the *angle of incidence* and is identified by the Greek letter *theta* θ. When the light ray reflects from the surface, it leaves the surface at an angle on the other side of the normal. This angle is called the *angle of reflection* and is identified by the Greek letter *phi* φ. *Within the same material, the angle of reflection of a light ray is always equal to the angle of incidence when the light ray is reflected from a smooth surface.*

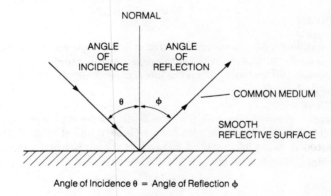

Angle of Incidence θ = Angle of Reflection φ

Figure 9-3. Light Reflecting from a Smooth Surface

APPLICATIONS OF PHOTOCOUPLED DATA TRANSMISSION SYSTEMS

9

Index of Refraction

The other important fact is that light travels at different speeds through different materials. The ratio of the speed of light in a vacuum to the speed of light in a material is referred to as the *index of refraction* of the material. *Figure 9-4* shows the effect on a light ray of two transparent materials with different indices of refraction. The effect is called *Snell's law of refraction* and is given in equation *1*.

$$n_1 \sin \theta = n_2 \sin \phi \qquad (1)$$

As an example, suppose the angle of incidence θ is equal to 30°. The light ray originates in air which has an index of refraction (n_1) of about 1. Assume it moves into a piece of glass with an index of refraction (n_2) of 1.4. Then the angle ϕ can be calculated as shown:

$$n_1 \sin \theta = n_2 \sin \phi$$
$$1 \sin 30 = 1.4 \sin \phi$$
$$\sin \phi = \frac{\sin 30}{1.4}$$
$$\phi = \sin^{-1}\left(\frac{\sin 30}{1.4}\right)$$
$$\phi = \sin^{-1}\left(\frac{0.5}{1.4}\right)$$
$$\phi = \sin^{-1}(0.357)$$
$$\phi = 20.9°$$

Recall that $\sin^{-1}$ or arcsin means that ϕ is the angle whose sine is the value of the ratio given. The ray will, therefore, enter the interface between air and glass at 30° off normal and exit the interface at 20.9° off normal.

Critical Angle

Now, suppose a light source is inside a plastic sheet with an index of refraction of 1.4, and the plastic sheet is surrounded by air as shown in *Figure 9-5*. Light leaving the LED strikes the surface of the plastic at an angle of θ and leaves at an angle of ϕ. Snell's law says that:

$$1.4 \sin \theta = 1 \sin \phi$$

Figure 9-5 shows that if the angle ϕ is 90° or larger, the light remains inside the plastic sheet. A computation using equation *1* shows that when ϕ is 90°, the angle θ is 45.58 degrees.

$$1.4 \sin \theta = 1 \sin 90°$$
$$1.4 \sin \theta = 1 \times 1$$
$$\sin \theta = \frac{1}{1.4}$$
$$\theta = \sin^{-1}(0.714)$$
$$\theta = 45.58°$$

9. Applications of Photocoupled Data Transmission Systems

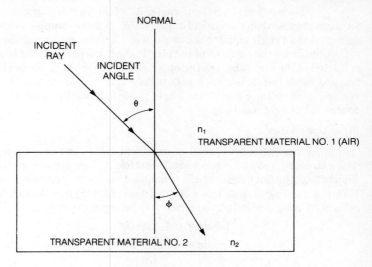

$$n_1 \times \sin\theta = n_2 \times \sin\phi$$
$$\text{WHEN } n_2 > n_1 \quad \sin\phi = \frac{n_1 \times \sin\theta}{n_2}$$

Figure 9-4. Snell's Law of Refraction

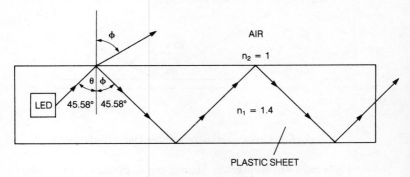

Figure 9-5. An Example of Snell's Law

APPLICATIONS OF PHOTOCOUPLED DATA TRANSMISSION SYSTEMS

9

This is called the *critical angle*. All of the light striking the surface at an angle greater than the critical angle (45.58° in this example) is internally refracted at an angle equal to the angle of incidence according to the rules of reflection. Since the surfaces of the plastic sheet are parallel, the beam is refracted at the same angle at the opposite surface so the refracted beam is trapped inside the sheet until it gets to the edge. As you can see, the parallel surfaces of the plastic sheet form a light guide to guide the beam from the source to the right hand edge.

<u>Fiber Optic Fiber</u>

Fiber optics are special cases of light guides and have been produced with extremely good transmission efficiencies. A fiber optics fiber is shown in *Figure 9-6*. The fiber consists of a cylindrical core with an index of refraction of 1.5 and a cladding with an index of refraction of 1.4. The critical angle can be calculated using Snell's law (equation *1*) to be approximately 69°.

$$1.5 \sin \theta = 1.4 \sin 90°$$
$$\sin \theta = \frac{1.4 \times 1}{1.5}$$
$$\theta = \sin^{-1}(0.933)$$
$$\theta = 68.96°$$

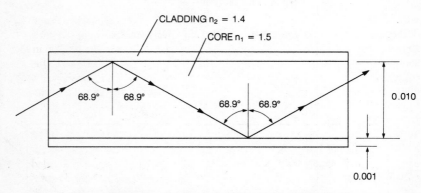

Figure 9-6. Cross Section of a Fiber Optic Light Guide

Light rays which enter the fiber at angles greater than 69° are internally refracted and stay inside the fiber until they reach the other end. Since bends in the fiber which are not abrupt do not significantly alter the efficiency, such fibers can be used to "pipe" light over long distances with little loss. Thus, by controlling the medium, light signals can be delivered from relatively low intensity sources over great distances to a detector. It may also be observed that light coming into the fiber from the side will penetrate through the fiber without being trapped inside, thus reducing unwanted signals.

9 APPLICATIONS OF PHOTOCOUPLED DATA TRANSMISSION SYSTEMS

APPLICATIONS OF DATA TRANSMISSION SYSTEMS USING AIR AS A MEDIUM

Infrared Remote Control

Infrared devices are being used widely to provide remote ON-OFF control and control of television volume and channel selection. The basic concept is simply to turn on an infrared light source in a hand-held controller and detect the light at the television to control the TV set. In actual application the job is not quite that easy, because the power source is small, unwanted light is present, and more than one operation is to be controlled. However, these problems can be solved effectively without great difficulty for distances up to about 30 feet.

Figure 9-7 is a block diagram of a typical system showing both the hand-held transmitter and the receiver in the TV set. To reduce the consumption of power in the battery powered hand-held unit, the LED may be pulsed at currents up to 1 ampere for short pulses and low duty cycles. (Duty cycle is the ratio of ON time to total ON-OFF time.) To avoid interference from unwanted sources of light, the LED may be pulsed at a frequency different from that of expected unwanted sources and the receiver circuit tuned to pass only the transmitted frequency. Fortunately, the solutions to these two problems are compatible.

One problem still remains; how can more than one task be performed with a remote control device which depends upon one light source and one detector? The three most likely ways are: (1) vary the brightness of the source; (2) vary the frequency of the pulses; or (3) code the information in the sending unit and decode it in the receiver. Changing the brightness of the source would not be effective since the unit must be capable of operating over a distance of up to 30 feet. Use of different frequencies to perform different functions could be effective using narrow bandpass filters in the receiver. However, this would limit the functions and increase the possibility of unwanted signals activating the receiver. Coding techniques, on the other hand, lend themselves well to solving all of the problems mentioned.

The block diagram of *Figure 9-7* contains the elements of a coded infrared remote control system. The receiver is matched to the frequency generator so the only signals passed to the decoder are those in the correct frequency range; therefore, unwanted signals have negligible effects. The keyboard provides inputs to the encoder which codes the information. The mixer passes the coded information in the form of bursts of signals from the frequency generator. The receiver accepts only the narrow band of frequency generated by the transmitter, decodes the signal, and performs the desired action that was selected by pressing the keys on the hand-held unit.

Applications of Photocoupled Data Transmission Systems

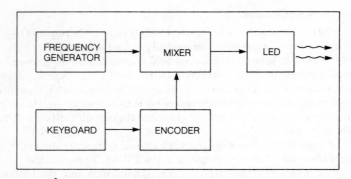

a. Hand-held Transmitter

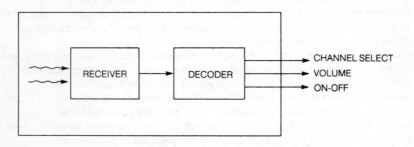

b. Receiver (In TV Set)

Figure 9-7. Infrared Remote Control

A typical transmitter using a special purpose integrated circuit is shown in *Figure 9-8*. This is a very compact circuit which uses only one 16-pin integrated circuit; three external resistors; five external capacitors; one crystal; one transistor; one VLED; and two infrared LEDs. It operates on 6 to 9 volts; is capable of providing control of up to 30 different functions; and provides a visible indication that it is transmitting. The clock frequency is 455kHz. The coded output signal is a 41kHz signal.

9. APPLICATIONS OF PHOTOCOUPLED DATA TRANSMISSION SYSTEMS

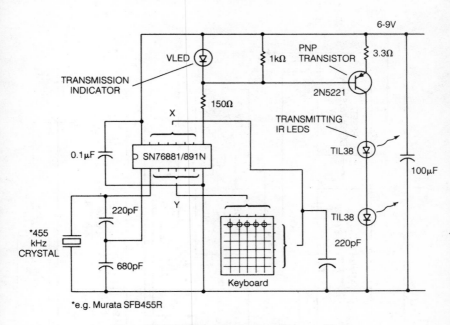

Figure 9-8. IR Remote Control Transmitter

The transmitter does not require an ON-OFF switch because the only power consumption in the standby mode is a leakage current of approximately 5 microamperes. *Figure 9-9* shows a block diagram of the SN76881 integrated circuit (IC). When a key is depressed, the power ON-OFF circuit is activated and a short time (about 10 milliseconds) is allowed to reset the strobe and code counters. Immediately after the turn-on delay, the strobe counter sends a 3-bit binary count which is changing sequentially to the 3-to-6 decoder to scan the 6 X lines of the keyboard. If a key is depressed, a signal from the X line goes out on the corresponding Y line from the keyboard to the 5-to-3 encoder which generates a unique 3-bit code for that Y line. These three bits along with the unique 3-bit strobe signal corresponding to the X line on which the key is depressed are loaded into a parallel in-serial out (PISO) shift register. The shift register outputs the data in serial form (one bit after another) to the output gate which modulates the IR LEDs. This is done twice under the control of the divide-by-two counter. The keyboard is then scanned again. If no keys are depressed, the circuit powers down to the standby mode. If the same key is still depressed or if another key has been depressed, the cycle repeats.

APPLICATIONS OF PHOTOCOUPLED DATA TRANSMISSION SYSTEMS 9

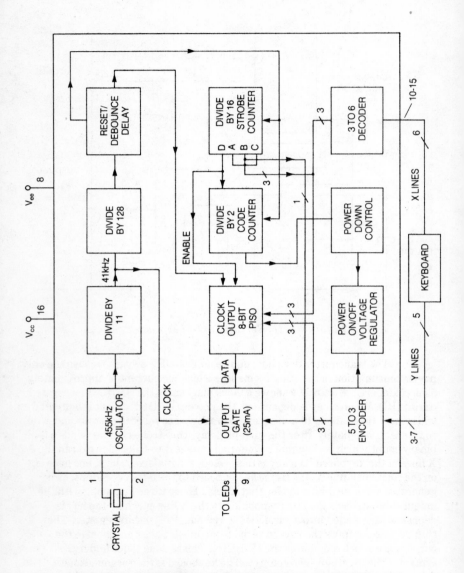

Figure 9-9. Block Diagram of the SN76881 Integrated Circuit

9 APPLICATIONS OF PHOTOCOUPLED DATA TRANSMISSION SYSTEMS

Figure 9-10 is a receiver which is compatible with the transmitter discussed. The light is detected by the TIL100 photodiode and is amplified by the preamplifier. The tuned input stage accepts the desired frequency while rejecting all others to minimize interference produced by electrical noise from the TV, lighting, or other sources. The output of the preamplifier goes to a differential-input-differential-output amplifier. The output of the differential amplifier is coupled directly to the demodulator.

The demodulator is comprised of a voltage controlled oscillator (VCO), a counter which divides the VCO frequency by 2, and phase detectors. The circuit bandwidth is set by the VCO, divider, and phase detector which are connected to form a phase locked loop (PLL). The PLL oscillates at a free-running frequency of 40kHz until an input signal with a frequency of 40kHz ± 10% is received. The PLL locks on this input signal and adjusts the output frequency and phase of the VCO and divider to match the output of the differential amplifier in frequency and phase. The demodulated pulses are reshaped by the pulse forming circuit. Then, the pulses are amplified and made available as an output at pin 4 of the IC.

An input signal to the receiver is a series of pulses of the proper frequency (40kHz) that contain the code set at the transmitter. The receiver demodulator produces a serial data output of bits like those originally produced at the transmitter. This data is fed to a serial in-parallel out (SIPO) shift register and decoded to activate the one of the thirty possible control functions that was selected by pressing the keyboard keys. The control function might be to turn off the TV, increase the volume, select a TV channel, etc.

Since the transmitter sends the coded data twice per transmission cycle, a good way to check for data errors is to store the first set of data in the receiver and compare it with the data received on the second transmission. This approach can be useful if the ambient light intensity varies over a wide range or if it is anticipated that the transmission path may be interrupted when the selection is made.

Wireless Audio Transmission

An interesting application of infrared photocoupled devices for wireless audio transmission is a communication system for astronauts. It is especially useful when they are working in space suits outside the spacecraft. The basic concept involves an IR LED, a VLED and a photodetector mounted on the helmet as shown in *Figure 9-11*. The infrared LED is used to transmit the audio signal while the VLED provides a visual indication of transmission to the receiving astronaut in case his helmet is oriented so that the detector does not receive the signal. The requirements of this communication system include low power consumption, small physical size, short range, and high reliability.

APPLICATIONS OF PHOTOCOUPLED DATA TRANSMISSION SYSTEMS

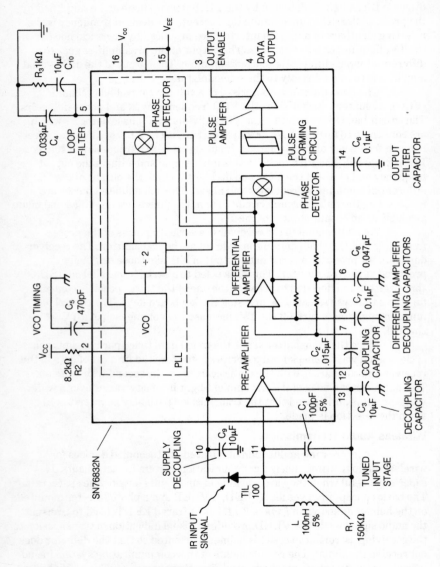

Figure 9-10. IR Remote Control Receiver

9

APPLICATIONS OF PHOTOCOUPLED DATA TRANSMISSION SYSTEMS

Figure 9-11. *Audio Communication Using LEDs*

A system meeting these specifications also would have application in high-noise work conditions such as those on construction sites. It would be useful for touring motorcycles where a combination of wind noise and helmets make communication between driver and passenger difficult. (Of course, the orientation of the transmitter would be different.) If the unit is used near conventional light sources which produce a 60Hz flicker, the receiver would pick up a 60Hz hum. This could be suppressed with a 60Hz active notch filter in the receiver or by modulating a carrier in the transmitter rather than transmitting audio directly. For the space, construction, and motorcycle applications, the 60Hz signal probably would not be a problem. Short range is not a problem either, because people are accustomed to the limited range of the physical voice when talking in normal conditions. Also, if several construction teams are working near each other, the short range is actually an advantage because the communication between members of one team does not interfere with other teams.

Selecting the Source and Detector

A transmitter and receiver which could be used in this application are shown in *Figure 9-12* and *9-13*. During design, we might consider the TIL31 and the TIL32 for the IR LEDs. The TIL32 produces 1.2mW of radiant energy at 20mA, while the TIL31 produces 6mW at 100mA. At first glance, it would seem that the TIL31 is the better choice since it produces quite a bit more radiant energy. However, the requirement of low power consumption makes the 20mA current level of the TIL32 the better choice because a check of data sheets shows that the TIL31 produces only 0.72mW at 20mA. The spectral response of both devices is the same and is compatible with a silicon phototransistor such as the TIL81.

APPLICATIONS OF PHOTOCOUPLED DATA TRANSMISSION SYSTEMS 9

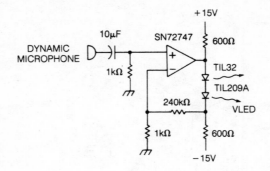

Figure 9-12. Audio Communication System Transmitter

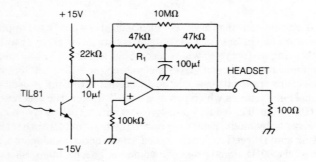

Figure 9-13. Audio Communication System Receiver

Another consideration is the viewing angle. The TIL31 has a viewing angle (half-intensity beam angle) of 10° while the TIL32 has a corresponding angle of 35°. For this application, the wider angle is probably the best choice, and again the TIL32 is favored. The TIL209A VLED is chosen to match the operating current of the TIL32.

For the detector, we could consider a silicon photodiode, avalanche photodiode, or phototransistor. For longer range, we might select an avalanche photodiode, but for the short-range requirement, a photodiode or phototransistor will provide adequate performance. Between the photodiode and phototransistor, a phototransistor is the better choice because of its higher sensitivity. If we compare two phototransistors for the application, we find that the spectral response of both the TIL78 and TIL81 will work well with the TIL32 LED. The viewing angle is 40° for the TIL78 and 10° for the TIL81. In

9. APPLICATIONS OF PHOTOCOUPLED DATA TRANSMISSION SYSTEMS

our application, the TIL78 would be the better choice based on viewing angle only. However, when the light current is considered, the data sheets show that the TIL78 produces only 0.5mA at 5 volts with a light intensity of $2mW/cm^2$ while the TIL81 produces 10mA under the same conditions. This specification favors the TIL81 by a large margin.

At this point, a decision must be made regarding whether the viewing angle or the sensitivity is more important. Since people usually turn to face a person speaking to them, the viewing angle for the detector may not be as important as sensitivity, so the TIL81 is selected.

The transmitter of *Figure 9-12* uses an operational amplifier and external circuitry which is designed to bias the LEDs at 20mA. The amplifier gain is set by the feedback network of the 240K and 1K resistor so that an audio input signal of 30mV rms from the microphone will fully modulate the LEDs. The receiver of *Figure 9-13* uses the TIL81 as a detector in the phototransistor mode. Since the capacitor from the collector of the TIL81 to the input of the operational amplifier has a low reactance at audio frequencies, the changes in light current generated by changes in the audio signal flow through this capacitor to the input of the operational amplifier. The input to the operational amplifier is held at almost zero volts or "virtual ground" by the feedback from its output. The signal current then flows through R_1 to ground through the 100µf capacitor. This produces a very high effective gain with the TIL81 acting as a signal current source and the operational amplifier acting as a current-to-voltage amplifier.

This system will have a range of approximately 20 meters. The range can be extended by adding lenses if desired, but ultimate range will be limited to between 50 and 75 meters unless the power levels are changed. The infrared LEDs can be operated in a pulsed mode with a peak current of one ampere as long as the average current and temperature are kept within specifications. This would extend the range, but a more complex circuit would be required.

APPLICATIONS OF OPTO-ISOLATORS

Specifications

An opto-isolator or opto-coupler has an infrared LED light source mounted in the same package as a silicon phototransistor with optical coupling between. The specifications for opto-couplers include specifications for the source and the detector, but the viewing angle is not specified because the optical path is determined by the manufacturer. Specifications that are important are the transfer curves as shown in *Figure 9-14a*, the input circuit to output circuit resistance (R_{IO}) and capacitance (C_{IO}), as well as the electrical isolation between input and output. Typical values for R_{IO} and C_{IO} are 100,000 megohms and 1 picofarad. Electrical isolation is in the order of 1 to 5 kilovolts. The transfer curves shown in *Figure 9-14a* can be used much as the output curves for a standard transistor. Package pin connections also are important and these are shown in *Figure 9-14b*.

APPLICATIONS OF PHOTOCOUPLED DATA TRANSMISSION SYSTEMS

9

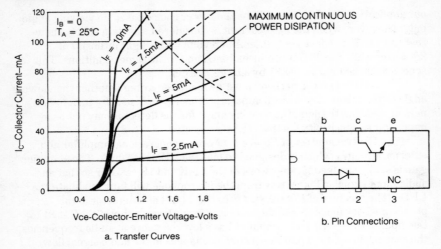

Figure 9-14. Opto-Coupler (Opto-Isolator) Specifications

Medical Field

One application of an opto-isolator in the medical field is illustrated in *Figure 9-15*. Here a patient has been wired with electrodes to measure medical data. If the patient is touching the metal table, or touches a lamp or other piece of equipment that is grounded, there may be a current path from ground point A through the patient to the probes or electrodes attached to his body and through the equipment to the equipment ground (B).

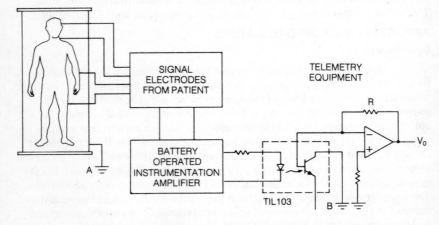

Figure 9-15. Isolation Amplifier for Medical Telemetry

9 APPLICATIONS OF PHOTOCOUPLED DATA TRANSMISSION SYSTEMS

Currents of only a few microamperes may be dangerous if they are allowed to flow in the body cavity, especially near the heart. To avoid this problem, the instrumentation amplifier is battery operated so it does not require a power-line ground. The opto-coupler then transmits the data to the monitoring equipment. The opto-coupler, in effect, places a resistance of approximately 100,000 megohms in series with any voltage which could cause a current through the patient. This would limit the current to 0.11 nanoamperes (1.1×10^{-10} ampere) even if a short-circuit or other fault applied 110 volts ac to the telemetry equipment. Notice that the phototransistor is used as a photodiode (emitter lead not connected) to provide current to the amplifier which is connected as a current-to-voltage amplifier.

Oil Field

An application of an opto-isolator in the oil field is illustrated in *Figure 9-16*. Loads on pump jacks, pump flow rates, and pump motor conditions are monitored and controlled by a remote computer system. The problem of grounding in this case is similar to the problem encountered in the medical example. Since the pumps are at different locations, different ground potentials could produce false signals or damaging currents through the computer input devices. Also, electrical noise generated by the motors can be transmitted through the cables to the computer and may interfere with the desired signal transmission. If lightning strikes in the oil field near the pumps or cables, large "spikes" of voltage may be induced in the cables and conducted to the computer where they could damage the system. Opto-isolators at the input to the computer system eliminate these problems.

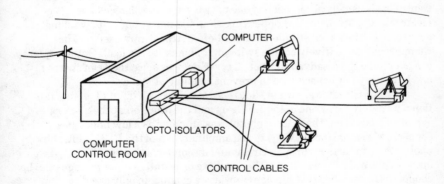

Figure 9-16. Oil Field Instrumentation

APPLICATIONS OF PHOTOCOUPLED
DATA TRANSMISSION SYSTEMS

9

Adding Analog Voltages

Another application for opto-couplers is adding voltages which have a common reference as shown in *Figure 9-17*. The terms K_1 and K_2 are constants that depend on the choice of R_1, R_2 and the transfer characteristics of the opto-couplers.

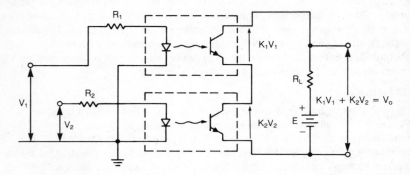

Figure 9-17. Opto-coupled Circuit to Add Two Voltages

Chopper Circuit

Still another application for an opto-coupler is in a chopper circuit. A chopper converts a dc signal or very low frequency ac signal as shown in *Figure 9-18a* to an ac signal so that high-gain, stable ac amplifiers may be used. This is usually accomplished one of two ways, mechanically or electronically. Mechanical choppers are slow and noisy, but fast switching electronic devices such as bipolar or field-effect transistors used for the chopper may capacitively couple switching "spikes" to the load. Opto-couplers can overcome the disadvantages of all these devices since they can switch fast and have very low capacitance to prevent coupling of spikes.

The circuit is shown in *Figure 9-18b*. The clock input (T) to the flip-flop alternately turns one of the LEDs on and the other off at the clock rate. When an LED is on, the mating phototransistor acts as a low impedance and connects the signal input (V_{IN}) to the amplifier. When the LED is off, the phototransistor acts as an open circuit and disconnects the signal input. Two opto-couplers are used so that either positive or negative input voltages can be chopped. (Remember that bipolar transistors conduct in only one direction.) When the input voltage is positive, the bottom phototransistor is the chopper and when the input voltage is negative, the top phototransistor is the chopper.

9. APPLICATIONS OF PHOTOCOUPLED DATA TRANSMISSION SYSTEMS

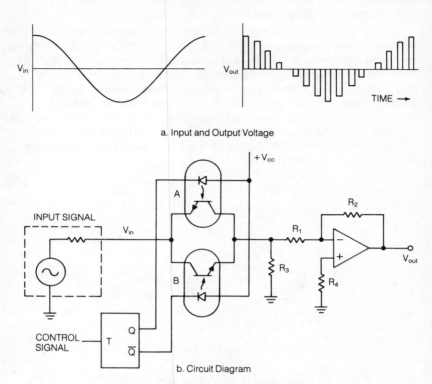

Figure 9-18. Chopper Circuit

APPLICATIONS OF FIBER OPTICS

There are many cases where light is to be directed in small concentrated areas. Fiber optics are ideal for this, because they can be used as "light pipes" to route light to selected concentrated points. An automobile instrument panel is one example. Directing light to selected places inside the human body so medical personnel can see to investigate medical problems or to perform operations is another example. However, the principal use of fiber optics will be in the communications field; not for the light itself but to use light as the carrier of other information.

One reason for this is the wide band width and another is the improvement in transmission efficiency. When optical fibers were first considered for communications use in 1968, the transmission losses were above 1000dB/Km (roughly 20% loss per meter). In 1970, Corning Glass Works produced a fiber optic cable several hundred meters long with losses under 20dB/Km (less than 0.5% loss per meter). At the present time, losses are approximately 2dB/Km (less than 0.05% loss per meter).

APPLICATIONS OF PHOTOCOUPLED DATA TRANSMISSION SYSTEMS

9

Characteristics of Fiber Optics

Figure 9-19 shows cross-sectional views of four types of optical fibers with the typical diameter ranging from 3 to 4 mils. In *Figure 9-19a* the fiber is unclad. n_o is the index of refraction of air and n_1 is the index of refraction of the fiber. These unclad fibers are not normally used because the optical characteristics change when they are supported and the transmission properties are disturbed. *Figure 9-19b* is a clad fiber with the index of refraction for the cladding (n_1) being greater than air (n_0) but less than the core (n_2). Because of the small inner core, this fiber is good for coherent light transmission (laser light) but not for incoherent sources (LED). The larger cladding is necessary primarily for strength.

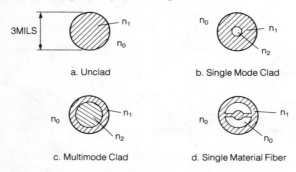

Figure 9-19. *Cross Section of Optical Fibers*

Figure 9-19c is a clad fiber which is called a multimode fiber and is used for incoherent light. It has a larger inner core. *Figure 9-19d* is a single material fiber in which the outer shell isolates the core from outside disturbances such as supports. All of these fibers can be coated with a thin film of optically lossy material to absorb any radiation which may scatter from the core to the cladding. This minimizes cross talk in multifiber bundles. (Cross talk is caused by an original fiber signal entering another fiber through the walls and then being transmitted along that fiber.)

The basic loss mechanisms in fibers are material absorption; material scattering; radiation due to bends; and cladding losses. Material absorption accounted for most of the 1000 dB/Km losses mentioned earlier. The most common material now in use is a fused silica core and fused silica cladding. A graph of attenuation versus wavelength for fused silica fiber is shown in *Figure 9-20*. Notice that losses are dependent on wavelength and specifically that the attenuation peaks at approximately 9600Å. This means that infrared LEDs which radiate at 9600Å should not be chosen as a source for this light fiber if optimum results are needed. Infrared LEDs radiating in the range of 7800Å to 9000Å should be selected to drive the fused silica optical fibers if incoherent light is to be used.

9 APPLICATIONS OF PHOTOCOUPLED DATA TRANSMISSION SYSTEMS

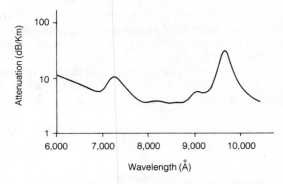

Figure 9-20. Attenuation as Function of Wavelength for Fused Silica Fiber

Advantages of Fiber Optics

Fiber optics have many advantages over common metallic transmission lines such as RG-58U and waveguides used in radar and microwave transmission. Compared to RG-58U and X-Band waveguide, the attenuation of 2dB/Km for fiber optics is less and the cost is significantly less. Fiber optics can also handle greater bandwidths than coaxial systems. This means that more channels of information can be handled. With laser sources the modulation bandwidth is in the gigahertz range, while the modulation bandwidth with IR LED sources is in the hundred megahertz range.

Small size (3 to 4 mils) and light weight are advantages in many applications. When the Navy replaced the conventional wiring used for communication between the computer and the avionics systems on the A7 aircraft, 224 feet of fiber optics replaced 1,900 feet of copper wire. The copper wire weighed 30 pounds compared to only 1.52 pounds for the fiber optics.

Another advantage is that fiber-optic cables are nonconductive and noninductive which means the cables will not couple electrical noise or interference from common sources as other cables do. For example, a lightning strike in the vicinity of a fiber optic cable will not be coupled by the cable to the transmitter or receiver. The cables are also immune to interference from radio transmitters or other electromagnetic sources.

Fiber optics also have disadvantages. The fiber optic cable cannot carry the dc current used in some communication systems for signaling. Separate wires must be run or a pulse technique used to provide the control. If a fiber optic cable breaks it must be spliced to be repaired. This can present a problem. The splice must be done very carefully so that all fibers are aligned to avoid losses in the splice.

APPLICATIONS OF PHOTOCOUPLED DATA TRANSMISSION SYSTEMS

9

Techniques of Data Transmission Using Fiber Optics

Fiber optic cable communications systems are comprised of a source of light (which may be an LED, a solid-state laser, or a semiconductor laser) coupled at one end of a fiber optic cable to a phototransistor detector connected to the other end of the cable. Good optical connectors are used at the interface between the light source and cable; between the cable and the detector; and at any splice in the cable. Standard multiplexing and demultiplexing techniques are used to interface to the light source and light detector as shown in *Figure 9-21*.

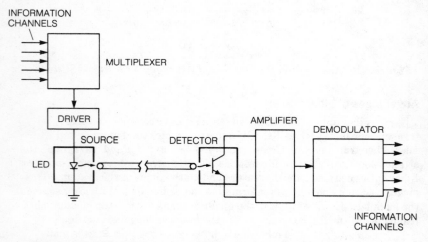

Figure 9-21. *Block Diagram of Fiber Optic Communications System*

The multiplexer combines the inputs from several channels of information into a single composite signal which is used to modulate the LED or semiconductor laser. The composite signal is transmitted through the fiber optic cable and converted to an electrical signal by the photodetector at the receiver. The demodulator separates the composite signal to reproduce the original channels of information.

For low data rates of less than 50 million bits per second (50 megabits) and for short distance applications, LEDs are suitable light sources. For high data rates of greater than 50 megabits per second, semiconductor lasers are the better choice for a light source. The number of channels depends on the type of data being transmitted. For telephone voice transmission, a data rate of 50 megabits per second allows 780 voice channels to be combined using time division multiplexing. These 780 voice channels can then be transmitted over one transmission link. A data rate of 150 megabits per second allows 2340 voice channels. A basic time division multiplexing system is shown in *Figure 9-22*.

9. Applications of Photocoupled Data Transmission Systems

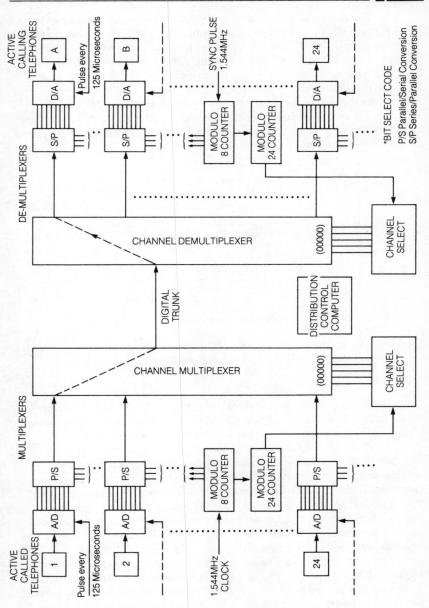

Figure 9-22. *Basic Time Division Multiplexing Hardware*
(Source: Understanding Communications Systems, D.L. Cannon and G. Luecke,
Texas Instruments: 1980)

APPLICATIONS OF PHOTOCOUPLED
DATA TRANSMISSION SYSTEMS

9

The A/D block is an analog-to-digital converter which samples the voice signal 8,000 times per second and generates an 8-bit code that represents the voice signal. The P/S block is a parallel-to-serial converter that changes the 8-bit code from parallel form to serial form. The P/S is controlled by the modulo 8 counter which selects each of the eight parallel bits in sequence to send one at a time to the channel multiplexer. The modulo 24 counter increments the channel select to the next channel, and the eight bits of the next channel are converted to serial form and placed in the sequence as shown in *Figure 9-23*. At the receiving end, the channels are separated by reversing this process in the demultiplexers.

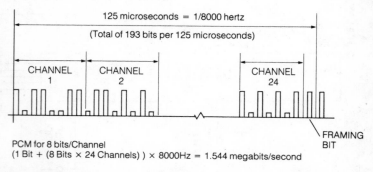

Figure 9-23. *Basic Time Division Multiplexing for 24 Channel System*
(Source: Understanding Communications Systems, D.L. Cannon and G. Luecke,
Texas Instruments: 1980)

Fiber optic transmitter-receiver sets are commercially available for data rates of 150 megabits per second for use with cable up to 5 kilometers long and with a bit error rate of less than one error in one billion bits.

These systems use a semiconductor double-heterostructure aluminum gallium arsenide diode laser. One of the problems with the semiconductor laser is that the output power varies more with changes in temperature than a LED. Therefore, feedback is used to stabilize the laser output power. As shown in *Figure 9-24*, the input modulation signal comes from the time division multiplexer to a buffer amplifier. The buffer output signal goes to an error amplifier and to the modulation current source for the semiconductor laser. A portion of the laser output is monitored by a photodiode which provides a sample of the transmitted signal to the error amplifier. The amplitude of the two signals is compared, and the output of the error amplifier is used to adjust the laser bias current to stabilize the output power of the laser.

9 APPLICATIONS OF PHOTOCOUPLED DATA TRANSMISSION SYSTEMS

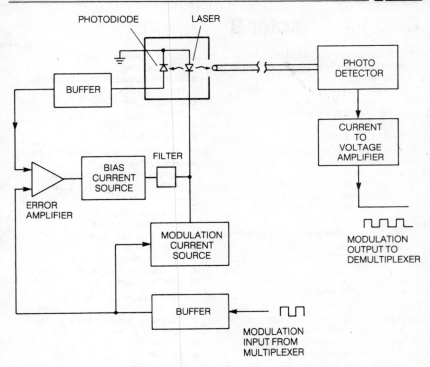

Figure 9-24. System to Modulate Laser Source

WHAT HAVE WE LEARNED?

1) Photocoupled data transmission systems send information from a source to a detector.
2) Changes in air temperature along the light's path can cause the light to "bend".
3) Materials that appear opaque to the human eye may be transparent to an optodetector.
4) The velocity of light and, therefore, the index of refraction is not the same for all materials.
5) Opto-couplers have high electrical isolation between input and output.
6) Fiber optics have a higher modulation bandwidth than coaxial systems.
7) Fiber optics cables have losses that depend on the wavelength of the light.
8) Fiber optics cables are nonconductive and noninductive. As a result, they do not couple common types of electrical noise.
9) Fiber optics cables must be spliced very carefully.
10) Common multiplexer techniques are used for fiber optic communications systems.

APPLICATIONS OF PHOTOCOUPLED
DATA TRANSMISSION SYSTEMS

9

Quiz for Chapter 9

1. A photocoupler (opto-isolator) has
   a. a controlled light path from source to detector.
   b. high resistance between the light source and light detector.
   c. high breakdown voltage between input and output.
   d. all of the above.

2. Fiber optic cables have losses because of
   a. temperature.
   b. ambient light.
   c. material absorption and scattering.
   d. none of the above.

3. When light reflects from a smooth surface the angle of reflection is
   a. greater than the angle of incidence.
   b. less than the angle of incidence.
   c. equal to the angle of incidence.
   d. none of the above.

4. When light travels from a medium with a high refraction index to a material with a lower refraction index, the angle of the light (with respect to the normal) is _____ when it leaves the interface.
   a. the same
   b. larger
   c. smaller
   d. none of the above

5. Infrared LEDs are popular sources for fiber optics because
   a. their wavelength is in a low loss band.
   b. they are inexpensive.
   c. they are fast.
   d. all of the above.

6. For long distance transmission with optical systems, the best transmission media is
   a. air.
   b. fiber optics.
   c. water.
   d. a vacuum.

7. Infrared remote control is popular because
   a. several channels can be controlled inexpensively.
   b. the light is visible to the human eye.
   c. the range is very long.
   d. none of the above.

8. Fiber optic cables use a cladding to
   a. keep light in the core.
   b. provide mechanical strength.
   c. electrically insulate the core.
   d. a. and b. above.

9. In some systems, both infrared and visible LEDs are used
   a. to produce more power.
   b. so that the operator will know that the system is operating.
   c. for temperature compensation.
   d. all of the above.

10. An opto-coupler is useful in a chopper circuit because
    a. it provides isolation between the input and detector.
    b. it is faster than mechanical relays.
    c. it has lower capacitance than FET switches.
    d. all of the above.

1. d, 2. c, 3. c, 4. b, 5. d, 6. d, 7. a, 8. d, 9. b, 10. d

10 APPLICATIONS OF LASERS

Applications of Lasers

ABOUT THIS CHAPTER

Previous chapters have dealt with a wide spectrum of light at many different frequencies and wavelengths. In this chapter, the discussion centers on devices that operate at single frequencies. Such devices are called lasers. The theory of operation; the types of lasers; and some applications will be covered.

WHAT IS A LASER?

A laser is a device which produces a special kind of light called coherent light. As a result, laser light stays concentrated in a very narrow intense beam. By contrast, light from incandescent and fluorescent sources is called incoherent because it scatters in all directions. Have you ever observed the way a large group of children scatter when released to the playground? That's the way incoherent light behaves. Have you ever watched a uniformed drill team marching in exact step and all moving in the same direction? That's the way coherent light behaves.

There are two types of coherence. *Temporal coherence* refers to how nearly monochromatic a light source is. Monochromatic means "one color"; that is, the light waves have the same frequency and wavelength. The higher the temporal coherence of a light source, the nearer it is to being purely monochromatic. *Spatial coherence* refers to how well all the individual waves in a beam of light move together. The high spatial coherence of laser light is what allows it to form such a narrow, intense beam. Laser light possesses both spatial and temporal coherence; although the temporal coherence of the light from some types of lasers is much better than that of other types.

CAUTION

Never look into a laser source or mirrored reflection. The concentrated energy in the laser beam can cause permanent damage to the eye.

APPLICATIONS OF LASERS   10

HOW DOES A LASER PRODUCE COHERENT LIGHT?

In order to understand how a laser produces coherent light, we must first discuss the emission and absorption of photons of light. Look at *Figure 10-1*. Suppose we have a single atom which can exist with two different states, E_1 and E_2. Let E_2 be the state with higher energy. If the atom is in state E_1, it is possible for it to absorb a photon with energy E_2-E_1. The atom will then be raised to energy state E_2 as shown in *Figure 10-1*, and the situation is referred to as absorption of a photon of light.

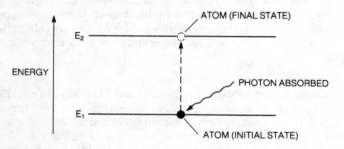

Figure 10-1. Absorption of a Photon

When an atom is already in energy state E_2, it is possible for it to drop back to energy state E_1 without external influence. In this case, the result is opposite to absorption; that is the atom emits or radiates a photon with energy E_2-E_1. This is called *spontaneous emission* of a photon, and it is illustrated in *Figure 10-2*.

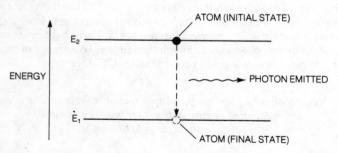

Figure 10-2. Spontaneous Emission of a Photon

10. APPLICATIONS OF LASERS

When the atom is in energy state E_2, there is another way for it to return to energy state E_1. Look at *Figure 10-3*. If a photon with energy E_2-E_1 is incident on (strikes) the atom while it is in energy state E_2, the incident photon can stimulate the atom to drop to energy state E_1 and emit a second photon with energy E_2-E_1. This situation, called *stimulated emission* of a photon is *the mechanism a laser uses to generate light*.

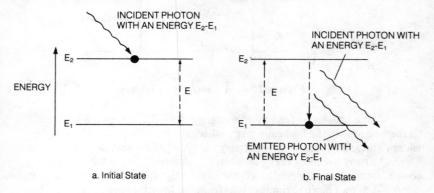

Figure 10-3. Stimulated Emission of a Photon

Multiple Energy Levels

Most lasers use material with atoms (or ions or molecules) that can be in three or four different energy states instead of two. Lasers which use three energy states are called three-level lasers and lasers which use four energy states are called four-level lasers.

The first laser built was a three-level laser which used the energy states of chromium ions in a ruby crystal. A diagram of these energy states is given in *Figure 10-4*. At room temperature, almost all of the chromium ions are in the lowest energy state.

The first step in producing laser light consists of moving a large proportion of the chromium ions from the lowest energy state E_1 into the highest energy state E_3. The process of adding energy for this purpose is called *pumping*. For the ruby laser, pumping is accomplished by exposing the ruby crystal to the intense flash from a powerful xenon flash tube. Chromium ions in the lowest state, E_1, absorb photons from the flash tube and are raised to the highest energy state, E_3.

In *Figure 10-4* there is a middle energy state, E_2. If the middle energy state were not present, the pumping process would result in half the chromium ions being in the low energy state, E_1, and half in the high energy state, E_3. In this case, if a photon with the appropriate energy entered the laser, it would be equally as likely to be absorbed as it would to stimulate the emission of a photon. For a laser to work, stimulated emission must be much more likely than absorption.

APPLICATIONS OF LASERS 10

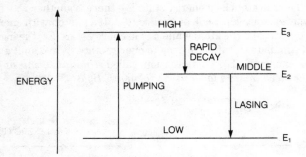

Figure 10-4. A Three-Level Laser

The presence of the middle energy state, E_2, solves the problem. Chromium ions in the highest energy state, E_3, rapidly drop to the middle energy state, E_2, and the excess energy is given off as heat to the ruby crystal. Thus, the ruby becomes very hot when it is pumped and usually must be cooled. As the pumping continues, the rapid emptying of the highest energy state, E_3, allows more chromium ions from the lowest energy state, E_1, to be raised to the highest energy state, E_3. At some point, more chromium ions will be in the middle energy state, E_2, than in the lowest energy state, E_1. This situation is referred to as population inversion. *Population inversion* means that more ions are in a high energy state (E_2 in *Figure 10-4*) than are in a low energy state (E_1 in *Figure 10-4*).

When a population inversion is present, a photon with the proper energy will be much more likely to stimulate the emission of another photon rather than to be absorbed. In fact, *one* photon *can stimulate* the emission of *many* other photons as it moves through the ruby crystal. All of these emitted photons will have approximately the same energy and the light will be essentially monochromatic. This process is called <u>L</u>ight <u>A</u>mplification by <u>S</u>timulated <u>E</u>mission of <u>R</u>adiation and "LASER" was the acronym coined for the name. (Because of popular usage, laser is now a word rather than an acronym.)

Ruby Laser

Figure 10-5 depicts a typical ruby laser setup which shows the xenon flash tube for pumping the laser, the ruby crystal, the tubes for the coolant, and two mirrors, one at each end of the ruby rod. (In the original ruby laser, these mirrors were actually part of the rod. They were formed by polishing the ends of the ruby rod until they were exactly parallel and then coating them with a reflective substance.) Emitted photons are reflected back and forth through the ruby rod by these mirrors. This causes even more photons to be emitted, and in addition, groups all the photons together as they bounce back and forth. This helps ensure spatial and temporal coherence.

10 APPLICATIONS OF LASERS

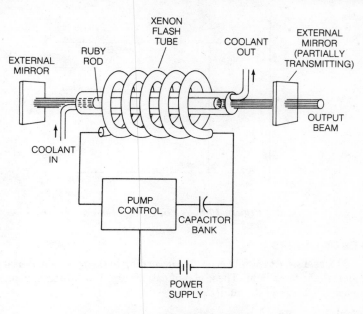

Figure 10-5. A Typical Ruby Laser

Notice that one of the mirrors is partially transmitting. When enough photons have been emitted by the excited chromium ions, the laser beam will have enough energy to escape through the partially transmitting mirror. This type of laser is usually pumped so that it emits pulses of laser light instead of a continuous beam. This allows the ruby rod to cool during the "OFF" time to prevent it from overheating.

Four-Level Laser

The energy levels in a four-level laser are shown in *Figure 10-6*. In this type of laser, the three upper energy states are essentially unpopulated at room temperature. Atoms (or ions) are pumped from the lowest state, E_1, to the highest state, E_4. Atoms in the highest state, E_4, decay rapidly to the next lower state E_3. Since energy state E_2 is essentially empty, a population inversion exists as soon as atoms drop into energy state E_3. Hence, laser action can occur more rapidly and with less pumping energy than in a three-level laser. Continuous wave operation, as opposed to pulse operation, is much easier to achieve with four-level lasers.

APPLICATIONS OF LASERS

10

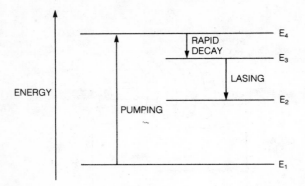

Figure 10-6. A Four-Level Laser

TYPES OF LASERS

Lasers are typically classified according to the type of material used to generate the laser light. These classifications are: 1) solid-state, 2) gas, 3) semiconductor, and 4) liquid.

Solid-State Lasers

The term solid-state is used in a different sense when applied to lasers. It refers to materials which have impurity ions in insulating host materials which are optically "pumped" for use as a laser. The term "solid-state" should not be confused with semiconductor when discussing lasers.

Solid-state lasers are made of ions of a transistion metal (chromium, manganese, cobalt, or nickel) or of rare earth elements (atomic numbers from 58 to 71) embedded in a solid host material made of crystal or glass. The ruby laser described above is one example of a solid-state laser. The ruby rod consists of chromium ions embedded in a sapphire crystal and produces light with a wavelength of 6,943 angstroms. The neodymium laser is another popular laser which consists of neodymium ions embedded in glass. Another common solid-state laser is the neodymium in yttrium aluminum garnet (YAG). The neodymium laser is a four-level laser and produces light with a wavelength of 10,600 angstrom. All of these solid-state lasers are "pumped" by optical excitation.

10 Applications of Lasers

Gas Lasers

Gas lasers use a gas contained in a glass tube. There are several gasses that may be used. Gas lasers are pumped by an external source of radio frequency energy and dc current. Spherical mirrors instead of flat mirrors are normally used. A diagram of a typical gas laser is given in *Figure 10-7*. The ends of the glass tube containing the gas have special windows to allow the laser light to pass in and out of the glass tube without being reflected. These windows are called Brewster windows because they are oriented at a special angle, termed the Brewster angle, which minimizes reflection.

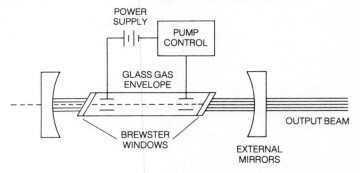

Figure 10-7. A Typical Gas Laser

Gas lasers can be divided into three main types depending on the nature of the energy transition involved in the laser action. Helium-neon and helium-cadmium lasers use the transitions between the energy states of non-ionized atoms. The argon and krypton lasers use transitions between ionized atomic energy states. The carbon dioxide laser uses energy levels involved in various states of molecular vibration and rotation.

Gas lasers can be made with highly coherent outputs. In addition, gas lasers are very efficient, and they provide laser light in a wide choice of wavelengths.

Semiconductor Lasers

Semiconductor lasers operate on a principle different from the three- and four-level lasers previously discussed. The semiconductor laser uses the properties of the junction between heavily doped layers of P and N type semiconductor material. If a large forward bias voltage is applied to this type of junction, a large number of free holes and electrons are created in the immediate vicinity of the junction and it is this that causes the population inversion. When a hole and electron pair collide and recombine, they no longer are free and a photon is produced. Since such a large number of holes and electrons are created in a very thin layer around the junction, there are very many collisions of holes and electrons. This produces the large number of photons needed for laser action.

APPLICATIONS OF LASERS   10

Figure 10-8 shows a typical gallium-arsenide laser. Two ends of the semiconductor crystal are cleaved or polished so they are parallel to form reflectors for the laser. The laser light is emitted from the junction through one of the parallel faces.

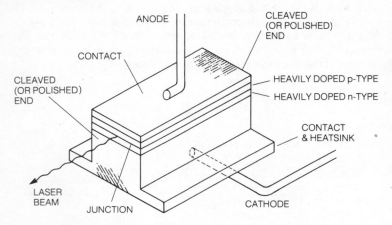

Figure 10-8. *A Typical Gallium Arsenide PN Junction Laser*

Due to the high forward bias current required for laser operation, early semiconductor junction lasers had to be operated in short pulses or cooled with liquid nitrogen if continuous operation was desired. The high forward current requirement can be reduced by using heterostructure semiconductor lasers. As can be seen in *Figure 10-9*, in a heterostructure laser, the PN junction is sandwiched between layers of material with different optical and dielectric properties. The material that shields the junction is typically aluminum gallium arsenide which has a lower index of refraction than gallium arsenide. This in effect traps the holes and electrons in the junction region and ensures the production of more photons for a given level of forward current. It also ensures that the photons are concentrated in the junction region as shown by the laser intensity profile in *Figure 10-9*. Heterojunction semiconductor lasers can be operated continuously at room temperature with no problems.

The production of laser light from a semiconductor laser is critically dependent on the forward bias current. Therefore, it is very easy to modulate the light from a semiconductor laser by simply varying the bias current. This characteristic and the fact that semiconductor lasers are relatively cheap, compact and efficient are making semiconductor lasers increasingly popular.

10. Applications of Lasers

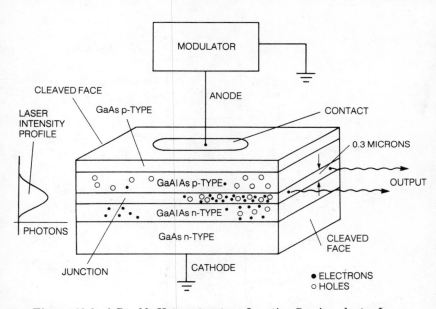

Figure 10-9. A Double Heterostructure Junction Semiconductor Laser

Liquid Lasers

Another class of laser uses a liquid in a glass tube as the laser medium. Out of the many types of liquids which have been used, the most popular have been organic dyes dissolved in a suitable solvent. A typical organic dye laser is shown in *Figure 10-10*. Organic dye lasers are usually pumped with either a flashlamp or another laser.

The best feature of organic dye lasers is they can be tuned so that the output light can be at different frequencies. This frequency tuning can be accomplished by changing the distance between the mirrors, changing the pumping energy, or by changing the concentration of the dye in the solvent. If one of the mirrors is replaced by a diffraction grating, as shown in *Figure 10-10*, narrowband tuning can be performed by tilting the grating. A good feature of all liquid lasers is that they can be cooled by recirculating the laser liquid.

APPLICATIONS OF LASERS   10

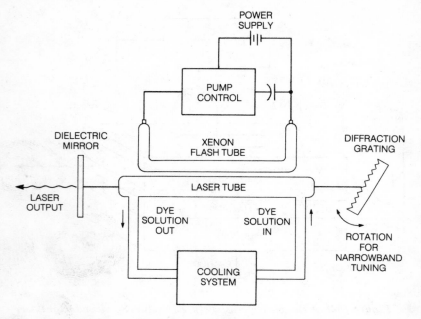

Figure 10-10. A Pulsed Organic Dye Laser

WHAT IS A MASER?

MASER was the acronym coined for Microwave Amplification by the Stimulated Emission of Radiation. (Because of popular usage, maser is now a word.) A maser is a device quite similar to a laser. In fact, masers came before lasers, and the research on masers contributed a great deal to the research leading to the laser. The main difference between masers and lasers is masers emit radiation in the microwave portion of the electromagnetic spectrum instead of in the infrared and visible light portions.

Ammonia Maser

The first maser was a two-level device which utilized transitions between two energy states of the ammonia molecule. A diagram of the ammonia maser is shown in *Figure 10-11*. The ammonia tank is used to provide a steady supply of ammonia molecules in both the high and low energy state. Although some ammonia molecules are always in a high energy state, the number of molecules in a high energy state may be increased by heating the ammonia. The electrostatic focuser separates the molecules by discarding the low energy ammonia molecules and passing the high energy molecules on to the resonant cavity.

10. Applications of Lasers

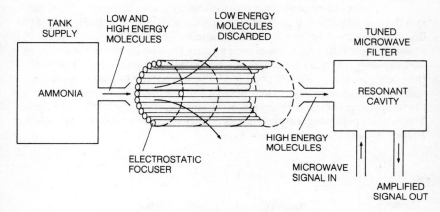

Figure 10-11. Ammonia Maser

The resonant cavity is essentially a tuned microwave filter. For the ammonia maser, the cavity is designed to resonate at 23,870 MHz because this is the frequency of the photons generated by the energy level which is equal to the difference in energy states between the low and high energy ammonia molecules. The resonant cavity serves the same purpose in a maser that the mirrors do in a laser; that is, to reinforce the buildup of photons.

When a low level microwave signal at 23,870 MHz from an external source is coupled into the resonant cavity, the high energy ammonia molecules are stimulated to return to the lower energy state. When they drop back to the lower state, additional microwave photons will be emitted. Once the emission process has been started, it will continue even if the external 23,870 MHz input signal is removed. Hence, the ammonia maser can be used as an oscillator. In fact, since the ammonia maser can't be tuned, the main use is as an oscillator, because if it is used as an amplifier, the bandwidth is much too small.

Ruby Maser

A maser which is more useful as an amplifier can be constructed by using a ruby crystal. The ruby maser is a three-level device which operates like a ruby laser, except that different energy states are used in the solid-state ruby maser.

APPLICATIONS OF LASERS 10

Figure 10-12 not only shows the energy states of the ruby laser but also the number of ions that exist in each state is indicated by the length of the line at each state. The longer the line, the more ions exist in that state. As shown in *Figure 10-12a*, at room temperature the energy states used in ruby for the maser action are all equally populated with chromium ions. But, if all the energy levels are equally populated, there is no way to pump chromium ions from one energy state to another. This problem can be solved by lowering the temperature of the ruby crystal by immersing it in a bath of liquid helium. The relative populations of the various energy levels after cooling are shown in *Figure 10-12b*.

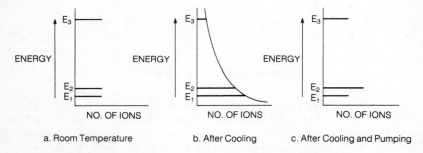

Figure 10-12. *Energy Level Populations in the Ruby Maser*

Finally, if the maser is pumped by a microwave signal containing photons with energy equal to the energy E_3-E_1, the populations are distributed as shown in *Figure 10-12c*. Notice that a population inversion exists between the middle and lower energy states. Now a small microwave signal which contains photons with energy E_2-E_1 applied to the resonant cavity will stimulate the emission of many more microwave photons. Thus, the signal will be amplified.

Figure 10-13 is a diagram of a typical ruby maser amplifier. The ruby crystal is placed in a resonant cavity that is immersed in a bath of liquid helium for cooling. There are two coupling inputs into the resonant cavity called ports. One port is used to provide the microwave signal for pumping the maser. The other port is used for coupling both the input signal to be amplified and the amplified output signal. This is accomplished by using a microwave circuit device called a circulator. The circulator allows the input signal to be coupled out as an amplified signal. The ruby maser can be tuned by applying an external magnetic field, as indicated by the permanent magnets in *Figure 10-13*. The magnetic field tunes the maser by shifting the energy states of the chromium ions.

10 APPLICATIONS OF LASERS

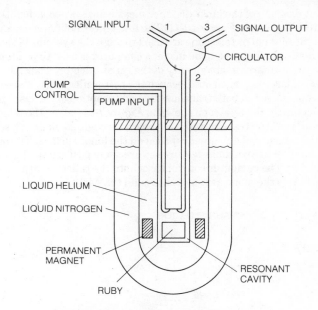

Figure 10-13. A Ruby Maser Amplifier

The main uses of masers have been as oscillators and low noise microwave amplifiers. Low noise refers to the fact that a maser amplifier has very little signal output (inherent noise) when there is no input signal. Such noise could cause interference with a very weak input signal. This is one reason why maser amplifiers have been used in radio astronomy and in radar receivers where weak input signals are common.

APPLICATIONS OF LASERS

Lunar Distance Measurements

An early application of lasers was to measure distance and this continues to be a useful area for laser application. Measuring the distance to the moon has been one of the most highly publicized uses of a laser for distance measurement. One of the items left on the moon by the first manned expedition was a retroreflector to aid in laser distance measurements from the earth. A retroreflector reflects radiant rays so the paths of the reflected rays are parallel to the incident rays. Retroreflectors are less sensitive to misalignment than plane mirrors. The original distance measurements were performed from Lick Observatory in California. Later, after larger retroreflectors had been placed on the moon in subsequent lunar landings, the distance measurements were moved to MacDonald Observatory in Texas.

APPLICATIONS OF LASERS 10

A diagram of the lunar distance measuring system at MacDonald Observatory is given in *Figure 10-14*. The low-power helium-neon alignment laser and the alignment telescope are used to align the system. When lunar distance measurements are to be made, a computer is used to position the telescope to the approximate vicinity of the retroreflector. Then, the powerful ruby ranging laser is turned on. Because the retroreflector is relatively small, the telescope is used in both transmission and reception of the laser beam. The small reflectors in the telescope shown in *Figure 10-14* allow the telescope operator to "see" the transmitted laser beam through the main telescope eyepiece as a small red dot superimposed on the lunar surface. The operator can then observe known lunar features to precisely position the beam to hit the retroreflector. The computer control locks onto the reflected beam so the telescope tracks the moon as it moves around the earth.

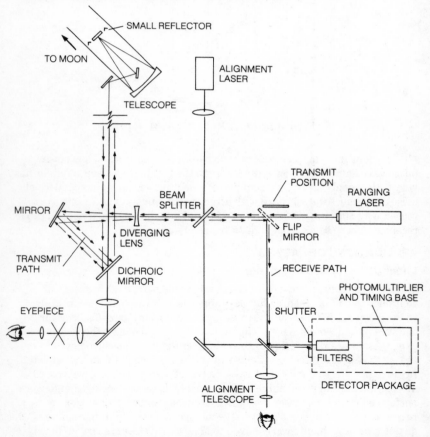

Figure 10-14. *The Lunar Ranging System at MacDonald Observatory*

10 APPLICATIONS OF LASERS

The basic concept in the distance measurement is to transmit a pulse of laser light to the retroreflector on the lunar surface. The reflected pulse is then received at the observatory, and the time interval between transmission and reception is accurately measured. Since light moves at a precisely known velocity, the distance the pulse traveled can also be accurately computed.

After the pulse has been transmitted, the flip mirror is flipped to the other position to convert the system to a receiver. Since the light pulse takes about 2.5 seconds for the round trip, there is enough time to move the mirror before the reflected pulse returns.

A great deal of filtering is performed to keep extraneous light from interfering with reception of the reflected pulse. In fact, the filtering is so effective that the system can even be used in the daytime.

Time intervals between the transmitted and received pulses can be measured to within about one nanosecond which corresponds to a distance of about three millimeters. In order to resolve such small distances as three millimeters out of a total distance of about half a million miles, the entire system must be very precisely aligned and calibrated.

The Geodimeter

The geodimeter uses modulated light from a helium-neon laser to measure distance. In order to use the geodimeter, a retroreflector must be placed at the point to which the distance is to be measured. To understand the operation of the geodimeter, refer to the block diagram in *Figure 10-15*. This system essentially measures the phase shift in the modulated signal that has been reflected from the distant point relative to a local reference signal. The modulating signal frequency has a certain wavelength associated with it. Since the distance the beam travels is to be divided by two (in order to find the distance from the geodimeter to the retroreflector), this system can measure distances which are multiples of wavelength/2 plus a fraction of wavelength/2. The accuracy of the fractional part of the measurement depends on the stability of the reference and modulating frequencies and on the ability of the phase detector to detect small phase differences.

In a typical application, the laser beam is modulated with a signal at the frequency f_T. The modulated laser beam must be aimed at the retroreflector. A special lens can be inserted to diverge the laser beam to provide a wider coverage at the target so the target can be located. The operator will "see" the retroreflector as a flash when the laser beam intercepts it. The special lens can then be removed and fine adjustments made to finish aiming the beam.

APPLICATIONS OF LASERS 10

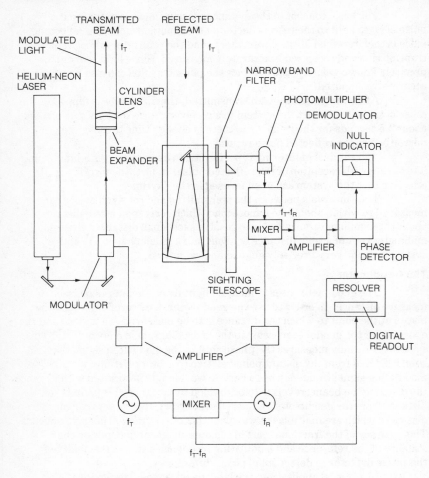

Figure 10-15. Block Diagram of a Geodimeter

The receiving telescope picks up the reflected beam and after filtering, the received beam is demodulated and the demodulated signal is mixed with a local reference signal at frequency f_R to produce a signal at frequency f_T-f_R. This signal is compared in the phase detector with a locally generated signal at the frequency f_T-f_R. A resolver with a digital readout is used to adjust the phase of the locally generated signal until the null indicator indicates the received and local signals are in phase. The distance measurement can then be determined from the digital readout. The geodimeter actually uses four different modulating frequencies and four

10 APPLICATIONS OF LASERS

different measurements are performed using the four different modulating frequencies. Then a series of equations are used to compute the distance measured. The geodimeter can measure distances up to 65 kilometers with an accuracy of ±5 millimeters.

Military

One of the most popular science fiction uses of the laser has been the death ray. With the advent of powerful lasers, a death ray no longer seems as far-fetched, but many problems still remain. Foremost among these is the high wattage power supplies required by such powerful lasers. These large, heavy power supplies could make transportation of such a laser very difficult. Another problem is that these devices only hit something directly in their line of sight so they must be aimed very very accurately to hit something at a great distance. The third problem is that atmospheric conditions such as rain or fog severely attenuate a laser beam. The fourth problem is that as a laser evaporates material, it creates a gas of destroyed material which tends to diffuse and block the laser beam, preventing it from doing any further damage.

As a result of these and other problems, death rays are still only a future possibility. However, in addition to destruction, there is some interest in using powerful earthbound laser beams to blind a spy satellite's optical sensor. There is also interest in using laser beams as weapons in space where weightlessness and lack of atmosphere remove some of the problems associated with earthbound laser systems.

So far, the biggest military use of lasers has been in tracking and ranging systems. The narrow beam width and immunity to electromagnetic interference of laser beam tracking and ranging systems give them advantages over conventional microwave tracking and ranging systems.

An example of a military laser beam tracking and ranging system is the laser system at White Sands Missile Range shown in *Figure 10-16*. The transmitting portion of the system, illustrated in *Figure 10-16a*, consists of the laser, a modulator, and a gimbal system to aim the modulated laser beam at the missile. The bender bimorphs are piezoelectric devices used to fine tune the beam positioning in the tracking system.

In the receiving portion of the system shown in *Figure 10-16b*, the laser beam reflected from the missile nose cone is received by a telescope which is concentric with the laser transmitter. Once the missile nose cone is located, the received image is processed by electronics to provide the target information needed for tracking to keep the transmitter pointed at the missile. The modulating signal in the reflected beam is phase compared to a local reference to provide distance information in much the same way as the geodimeter discussed earlier. Velocity of the target is obtained from the rate of change of distance. In all of these applications, the principles are very similar to RADAR (R̲Adio D̲etection A̲nd R̲anging), except that light frequencies are used instead of radio frequencies in the microwave region of the spectrum.

APPLICATIONS OF LASERS 10

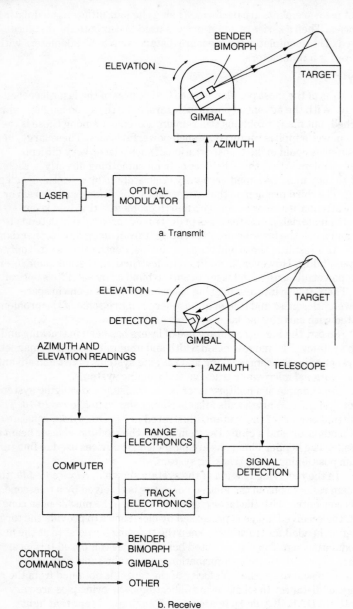

Figure 10-16. Ranging and Tracking System at White Sands Missile Range

10 APPLICATIONS OF LASERS

Medical

Another area in which lasers have proved to be very useful is surgery. The fact that a laser beam can be focused to a very small, intense spot has led to the use of lasers in reattaching detached retinas in the eye. A device called a laser photocoagulator is designed for this type of surgery. From the block diagram in *Figure 10-17*, it can be seen that the laser photocoagulator consists of a laser, optics to direct and focus the beam in the patient's retina, and optics to permit the doctor to aim the laser beam at the correct spot. The amount of energy delivered to the spot in the patient's retina can be varied as needed and a focus control permits adjustment of the spot size.

The laser photocoagulator has safety features to ensure the laser doesn't accidentally fire, and to protect the doctor's eye from the laser light.

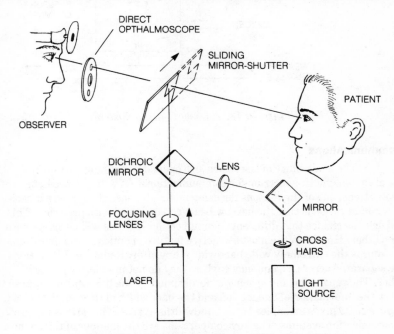

Figure 10-17. *Diagram of a Laser Photocoagulator*

Manufacturing

Lasers are also used for cutting and welding. The same features which make lasers useful in retina surgery allow them to make very small and precise cuts in virtually any type of material, no matter how hard or how brittle. *Figure 10-18* shows the schematic diagram of a laser cutting system. This particular system uses pressurized oxygen and a focused carbon dioxide laser beam to produce efficient cutting.

APPLICATIONS OF LASERS **10**

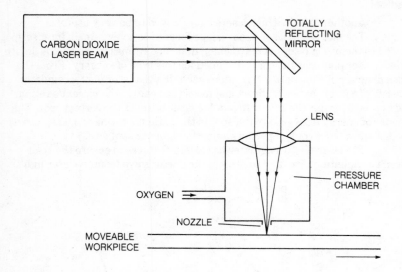

Figure 10-18. A Laser Cutting System

Communications

The availability of lasers has also opened a vast new portion of the electromagnetic spectrum for communications. As with the previous applications, one of the reasons for using modulated laser beams as a means of communication is that the narrow beam can be aimed with precision. This reduces the chance that different communication systems will interfere with each other. However, the most attractive feature of optical communications systems is the extremely wide bandwidth they offer. Radio and TV stations are separated across a frequency band so they do not interfere with each other. The separation and spacing is determined by the bandwidth required to carry the information of one station and the bandwidth of the complete band. *Figure 10-19* gives examples of the bandwidths available in various types of communication systems. The key comparisons are the number of telephone and TV channels. Notice that optical systems, even with a bandwidth of only 0.1% of the carrier frequency, can handle 1000 times more telephone channels and 1000 times more TV channels than the frequency bands through the microwave range. This enormous bandwidth capability makes optical communication systems very attractive in telephone links where many channels of voice and data communications are transmitted at the same time on one carrier.

10 APPLICATIONS OF LASERS

Band	Frequency Range	Usable Bandwidth	Approximate Number of Telephone Channels	Approximate Number of of TV Channels
LW	30kHz-300kHz	10%	3	-
MW	300kHz-3MHz	10%	25	-
SW	3MHz-30MHz	10%	200	-
VHF	30MHz-300MHz	10%	4000	1 to 4
UHF	300MHz-3GHz	10%	10,000	10
Microwave	3GHz-1THz	10%	100,000	100
Optical	5THz-1000THz	0.1%	10^8	10^5

Figure 10-19. Telephone and TV Channels Capacity Onto One Carrier in Various Frequency Bands

The main problem with optical communications is to find a proper transmission medium for the optical signal. There is not much of a problem in space, but earthbound optical communication systems can be adversely affected by many atmospheric conditions. In addition, the fact that the beam is very narrow and directional demands line-of-sight transmission through the atmosphere. This makes transmission very difficult and expensive in a mountainous environment or in a large city with many tall buildings.

We learned in Chapter 9 that fiber optics technology provides a transmission medium which solves these problems. A fiber optic guides light from one end to the other end of the fiber with very little loss. The fibers are flexible so they can be curved to guide light around corners. Bundles of fibers carrying many channels of communication can be buried underground in conduits just as conventional underground wire communications systems but requiring only a fraction of the space.

Figure 10-20 shows a simple fiber optics communications link. The light source is a laser. Information to be transmitted is in the form of digital pulses which modulate the light beam by turning it ON while a pulse is present and turning it OFF when the pulse is not present. The photodiode detector feeds the receiver. This system is dependent on the type laser for cost; is simpler than one where analog signal modulation is required; and can be made to operate at very high speed for high bit-rate digital transmission.

APPLICATIONS OF LASERS **10**

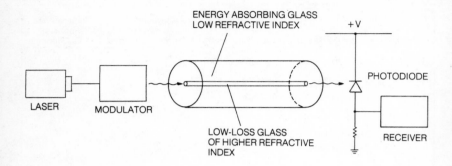

Figure 10-20. *Optical Fiber Link*

Video Disks

One of the most recent applications of lasers has been as part of the mechanism for recording and playing video disks. A video disk looks similar to a phonograph record and it contains the video, color, and audio portions of a television program. Video disks can carry two channels of high fidelity audio; therefore, stereophonic sound can be recorded.

Figure 10-21 is an operational diagram of a video disk recording system. In the recording system, the modulator gates the helium-cadmium laser beam according to the variation of pulses which contain the television program coded signal and the laser beam exposes the photoresist coating on the disk. (Photoresist is a material whose physical properties are changed by exposure to light. Areas exposed to light can be etched away when washed in an appropriate solution, but the unexposed areas cannot.) The areas of the photoresist that were exposed to the laser light are washed off while the unexposed areas stay on. This leaves pits in the disk surface as shown in *Figure 10-22*. The length of the pits and the distance between the pits correspond to the variations in the modulated laser beam containing the program. A thin nickel coating is applied to the developed master disk, and then evaporated metal is used to build up the thickness of the metal layer. The nickel coated master can then be used to produce copies for playing.

10. Applications of Lasers

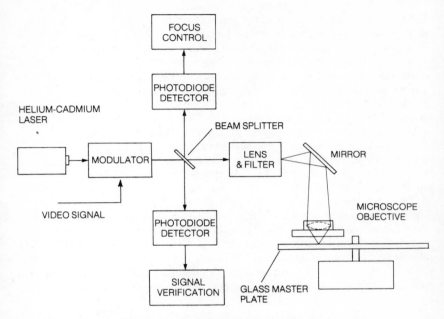

Figure 10-21. A Typical Video Disk Recording System

Figure 10-22. Video Disk Pit Structure Magnified 2700 Times
(Courtesy of Zenith Radio Coporation)

APPLICATIONS OF LASERS 10

The recorded program can be played on a video disk player connected to a television receiver. Each of the approximately 50,000 tracks on the video disk contains one complete television picture or frame. Each track is identified by a unique digital code so it is possible to randomly access any track within five to ten seconds. In the freeze-frame mode, each track is continuously replayed by having the read head jump back to the beginning of the track when it reaches the end of the track. This mode of playing a disk allows any frame to be leisurely inspected. Hence, the video disk has a great deal of potential as an information storage device. The video disk may eventually replace the magnetic disk for mass storage of digital data for computers. Experiments have demonstrated that a video disk has much greater storage capacity for digital data than a comparable size magnetic disk.

A diagram of a typical optical pickup for video disks is given in *Figure 10-23*. The system consists of a helium-neon laser with a light wavelength of 6,328 angstroms; a beam splitter; a photodiode detector; a mirror for radial tracking; and a microscope objective mounted in a movement which allows focus tracking of the video disk. The light from the laser is focused by the microscope objective to form a spot on the disk which is comparable to the size of a pit on the disk and the photodiode senses the laser light reflected from the disk. When the beam passes over a pit, the intensity of the light reflected from the disk diminishes and the output current of the photodiode drops.

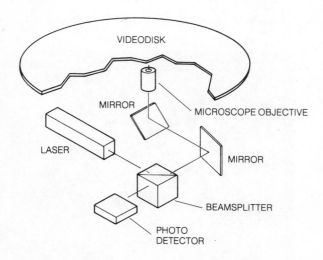

Figure 10-23. Typical Optical Pickup

10. Applications of Lasers

An optical feedback system is used to ensure that the laser beam remains focused on the disk despite variations in the surface of the disk. The radial tracking system also uses an optical feedback system to keep the laser beam centered over the track.

The surface of the video disk is covered with a layer of transparent material to protect the pits from damage. In addition, since the laser beam is focused to a point beneath the transparent layer; smudges, dust, and scratches in the transparent layer are out of focus so they don't affect the quality of the program recorded in the disk. Since the optical reading head never actually touches the disk, and the protective coating reduces the chance of handling damage, the lifetime of a video disk using a laser system should be very long.

High-speed Printer

High-speed printers provide another area for laser application. The Xerox 9700 laser printer (*Figure 10-24*) can achieve speeds up to 18,000 lines per minute using a modulated laser beam scanner. In this system, a helium-cadmium laser is modulated with an input signal coded with information that is to be printed is used to expose the surface of an electrically charged belt. Effectively, the charged belt can be considered as a matrix of capacitors with each one being charged to a level determined by the signal level as the beam scans. The exposed belt is then developed and the image is transferred to the copy paper to print it.

Figure 10-24. Xerox 9700 Printer
(Courtesy of Xerox Corporation)

APPLICATIONS OF LASERS 10

A diagram of the Xerox 9700 printer optics is given in *Figure 10-25*. One scan across one line is shown in the figure. The scan time is determined by the speed of rotation, 10,000 rpm, of the polygonal mirror. Since the angle of incidence of the laser beam on one of the flat mirrors of the polygonal mirror continuously changes, the reflected beam is swept across the surface of the charged belt. The belt is located in the focal plane of the cylindrical correction lens and the belt is charged according to the input information. The cylindrical correction lens corrects for small misalignments in the locations of the flat mirrors on the polygon. The other lenses in the optical system focus the beam to produce an essentially round spot of light on the belt. The start-of-scan and end-of-scan detectors synchronize the modulating information with the position of the beam. The modulating information is digital, so that both pictures and text can be sent as one continuous modulating signal stream. In addition, it is possible to use computer controlled character generation and graphics to generate any desired form.

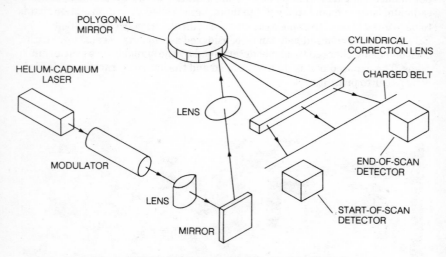

Figure 10-25. High-speed Printer Optical System

Holograms

One of the most publicized uses of lasers has been in the production of holograms. In essence, a hologram records on a piece of photographic film the information needed to reconstruct the original light waves reflected from an object illuminated with a laser. When a similar laser is shined through the developed hologram, the original light waves are reconstructed, and the viewer sees an image of the original object. Since the light waves the viewer of the hologram is seeing can't be distinguished from the original light waves

10. APPLICATIONS OF LASERS

reflected from the object, the image appears to be three-dimensional. It is even possible to look around and behind objects in the reconstructed image. The high coherence of laser light is what makes such three-dimensional holograms possible.

Figure 10-26a shows what is involved in recording a hologram. A laser is used to illuminate the subject of the hologram (object beam) and the same laser also illuminates the photographic plate. The direct beam from the laser to the plate is referred to as the reference beam. The reference beam and the light waves reflected from the subject interact to form a pattern of interference fringes on the holographic plate. The interference pattern shown on the plate of *Figure 10-26b* contains all the information necessary to reconstruct the light waves originally reflected from the object. In fact, every point on the holographic plate contains all the information necessary to reconstruct the original light waves. Even if only a small portion of the total hologram is cut off and used to reconstruct the original image, every detail will be present; however, the image won't be as bright.

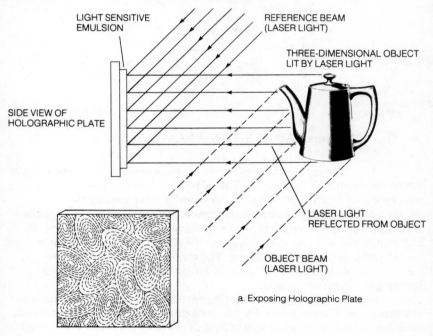

Figure 10-26. Recording a Hologram

APPLICATIONS OF LASERS 10

Figure 10-27 shows a system where the reference beam and the illumination of the subject are obtained from a single laser beam. When recording a hologram, it is very important that all the parts of the holographic recording system remain stationary. Holographic systems are usually placed on a special table desiged to keep vibrations to a minimum.

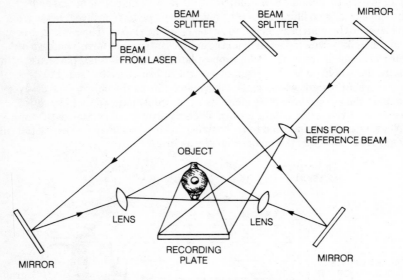

Figure 10-27. *A Holographic Recording System*

Once a hologram has been recorded and developed, the original image can be reconstructed. The basic technique is illustrated in *Figure 10-28*. A reference laser beam illuminates the developed plate containing the interference fringes which form the hologram. A viewer looking through the holographic plate will see the three-dimensional image. If the viewer moves and changes the angle his eye makes with the plate, the image will appear to shift just as if it were the real item. The viewer is actually seeing the reconstructed light waves caused by the holographic interference fringes breaking up the reference beam.

Apart from being an interesting novelty, holograms have a number of potentially useful applications. By using holographic techniques, variations as small as a quarter of a wavelength of light can be detected even in very rough or textured surfaces.

10. Applications of Lasers

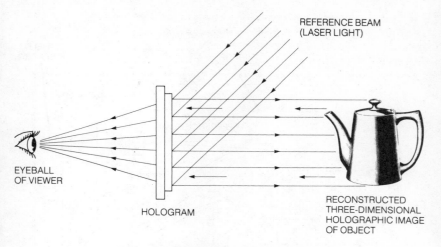

Figure 10-28. Viewing a Hologram

Tire Analyzer System

The tire analyzer system shown in *Figure 10-29* uses holograms to find defects in tires. The mirrors on either side of the tire are used to provide a simultaneous image of both sidewalls and the tread. A hologram is made of the uninflated tire. The tire is then inflated to a moderate pressure. The hologram of the uninflated tire is played back using the laser as a reference beam and the laser beam also illuminates the inflated tire. The holographic image of the uninflated tire and the real image of the inflated tire are superimposed (placed on top of each other). If the tire surfaces in the two images differ by a quarter wavelength of the laser light or more, interference fringes are generated on the superimposed images. An example of such an image is shown in *Figure 10-30*. This particular tire has two defects which show up as slight swellings when the tire is inflated. These very small swellings are completely invisible to the naked eye since they are only a few wavelengths of light in height, but they are shown very clearly by the tire analyzer system.

This is just one of the many applications of holography. Holograms have been used in character recognition, microscopy, and data storage. Proposals have been made for a video disk system which would store the video and audio signals as holograms on the disk.

APPLICATIONS OF LASERS 10

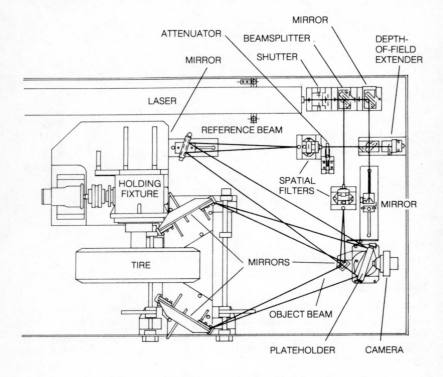

Figure 10-29. Diagram of Holographic Tire Analyzer

SUMMARY

In this chapter, we have seen a few applications of the laser. However, the laser is already being used in many others and it has potential for even more uses. Laser systems have been proposed to provide standards of frequency and length. Lasers are used in spectroscopy, both for providing contamination free samples for spectrographic analysis and as a tool in Ramon spectroscopy. Lasers are being used more and more in supermarket bar code readers. They are used in automobile manufacturing to code engine parts with bar codes and to read the bar codes to identify correct parts at an assembly line point. Techniques have been developed in image processing which use laser systems to restore a degraded image. With the continuing increase in the quality and availability of lasers, the list of applications is sure to keep growing rapidly.

10 APPLICATIONS OF LASERS

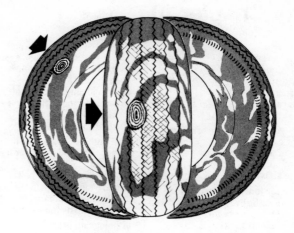

Figure 10-30. Defective Tire Using a Holographic Tire Analyzer

WHAT HAVE WE LEARNED?

1) Lasers are sources of coherent light.
2) How close a light source is to monochromatic (single frequency) is temporal coherence. How close individual waves making up a beam of light are in step is referred to as spatial coherence.
3) Lasers are pumped to produce a population inversion. When a population inversion exits, laser light is produced by the stimulated emission of photons.
4) The principal classes of lasers are solid-state, gas, semiconductor, and liquid.
5) Masers are devices which emit coherent microwave radiation by the same mechanism lasers used to produce coherent light.
6) Coherent radiation from a laser is very dangerous to the eye. A laser beam should never be directly viewed and even reflections of laser beams can be very dangerous.
7) Among the many applications of lasers are: measuring distance, tracking targets, communications, video disk systems, high-speed printers, cutting, welding, retinal surgery, spectroscopy, bar code readers, and holography.
8) A hologram is a recorded set of interference fringes which contain all the information necessary to reconstruct the light waves reflected from an object illuminated with laser light. When the hologram is played back using a laser beam, the reconstructed light waves cause the image to have all the visual properties of the original object.

APPLICATIONS OF LASERS 10

Quiz for Chapter 10

1. Light from the sun is
   a. spatially coherent.
   b. temporally coherent.
   c. incoherent.
   d. all of the above.

2. Gas lasers are
   a. available in a wide range of colors.
   b. efficient.
   c. capable of providing highly coherent outputs.
   d. all of the above.

3. Semiconductor junction lasers produce laser light by
   a. non-ionized atomic transitions.
   b. ionized atomic transitions.
   c. transitions between different states of molecular vibration and rotation.
   d. holes and electrons recombining in the region of the junction.

4. Organic dye lasers are useful because they can be easily
   a. tuned.
   b. pumped.
   c. modulated.
   d. understood.

5. The ammonia maser is useful mainly as
   a. an oscillator.
   b. a tunable amplifier.
   c. a low noise amplifier.
   d. a filter.

6. Laser systems use modulated signals to measure distance by performing
   a. phase comparison.
   b. amplitude comparison.
   c. time comparison.
   d. none of the above.

7. Communications systems using light beams have
   a. no atmospheric problems.
   b. wide bandwidths.
   c. no way of transmitting light around obstructions.
   d. very limited application.

8. Laser based video disk systems
   a. provide quick random access to any track.
   b. provide high fidelity stereo sound.
   c. are immune to dust and smudges in the disk.
   d. all of the above.

9. The high-speed laser printer discussed in this chapter uses the laser to
   a. expose a charged belt.
   b. burn small holes in the paper.
   c. develop the image.
   d. transfer the image to paper.

10. A hologram is a
    a. three dimensional picture.
    b. two dimensional picture.
    c. pattern of interference fringes.
    d. group of light waves.

1. c, 2. d, 3. d, 4. a, 5. a, 6. a, 7. b, 8. d, 9. a, 10. c

G Glossary

Glossary

Angstrom (Å): A wavelength value equal to 1/10,000 micron or 10^{-10} meter.

Avalanche photodiode (APD): A photodiode designed to take advantage of avalanche multiplication of photocurrent.

Beam (also ray): A slender shaft or stream of light or other radiation.

Brewster window: Glass windows placed at exact angles on the ends of a laser tube.

Cladding: A sheathing or cover of a lower refractive index material intimately in contact with the core of a higher refractive index material. It provides optical insulation and protection to the total reflection interface.

Coherence: Light waves which are in-step, in-phase, and monochromatic.

Core: The high refractive index central material of an optical fiber through which light is propagated.

Coupling loss: The total optical power loss within a junction, expressed in decibels, attributed to the termination of the optical conductor.

Critical angle: The maximum angle at which light can be propagated within a fiber.

Dark current (I_D): The current that flows through a photosensitive device in the dark condition.

DC fan-out: A term characterizing the driving capability of an output; specifically, the number of inputs of a specified type that can be driven simultaneously by the output.

Delay time (t_d): The time interval from the point at which the leading edge of the input pulse has reached 10% of its maximum amplitude to the point at which the leading edge of the output pulse has reached 10% of its maximum amplitude.

Detector: A device that converts optical energy to electrical energy.

Diffraction: Breaking up of light rays into bands when a ray is deflected by an opaque object or slit.

Diffusion: A scattering of light rays. A softening of light.

Electroluminescence: Electrical energy application to light-sensitive material causing emission of light (radiant) energy.

Fall time (t_f): The time duration during which the trailing edge of a pulse is decreasing from 90% to 10% of its maximum amplitude.

Fiber: A single discrete optical transmission element usually comprised of a fiber core and a fiber cladding.

Fiber cable: A cable composed of a single fiber or fiber bundle, strength members, and a cable jacket. The bundle may be a consolidated group of single fibers used to transmit a single optical signal.

Fiber optics (FO): A general term used to describe the function where electrical energy is converted to optical energy, then transmitted to another location through optical transmission fibers, and converted back to electrical energy.

Footcandle: Measure of intensity of illumination at all points of an illuminated surface one foot from a one candela source.

GLOSSARY

Forward voltage (V_F): The voltage across a semiconductor diode associated with the flow of forward current. The p-region is at a positive potential with respect to the n-region.

Frequency: The number of times a wave occurs in a period of time.

Hertz (Hz): Cycles per second of a frequency (10^3 is Kilo, 10^6 is Mega, 10^9 is Giga).

Hexadecimal display: A solid-state display capable of exhibiting numbers 0 through 9 and alpha characters A through F.

Illumination (E_v): The luminous flux density incident on a surface; the ratio of flux to area of illuminated surface.

Incoherent: Light waves that are not in-step, not in-phase, and nondirectional.

Index of refraction: The ratio of the speed of light in a vacuum to the speed of light in a material.

Infrared light-emitting diode: An optoelectronic device containing a semiconductor PN junction which emits radiant energy in the 0.78μm to 100μm wavelength region when forward-biased. This band of light wavelengths is too long for response by a human eye.

Irradiance (H or E_e): The radiant flux density incident on a surface; the ratio of flux to area of irradiated surface.

Laser: Light Amplification by Stimulated Emission of Radiation.

Light: Radiant energy transmitted by wave motion with wavelengths from about 0.3μm to 30μm. These wavelengths include invisible light, such as ultraviolet and infrared, and visible light.

Light current (I_L): The current that flows through a photosensitive device, such as a phototransistor or a photodiode, when it is exposed to illumination or irradiance.

Light-emitting diode (LED): See infrared-light-emitting diode and visible-light-emitting diode.

Lumen: Measure of luminous flux of one square foot of spherical surface one foot from a one candela source.

Luminance (L) (photometric brightness): The luminous intensity of a surface in a given direction per unit of projected area of the surface as viewed from that direction.

Luminous intensity (I_v): Luminous flux per unit solid angle in a given direction.

Maser: Microwave Amplification by Stimulated Emission of Radiation.

Micron: A wavelength value equal to 10,000 angstroms or 10^{-6} meter (1μm).

Microwave: Electromagnetic waves extending from 300 MHz to 300,000 MHz.

Monochromatic: Of, having, or consisting of one color.

Opaque: Material that blocks light rays.

Optical axis: A line about which the radiant energy pattern is centered; usually perpendicular to the active area.

Optical Spectrum: That part of the electromagnetic spectrum that includes infrared, visible, and ultraviolet wavelengths.

G GLOSSARY

Optically-coupled isolator (optical coupler): An optoelectronic device consisting of a photoemissive device and a photosensitive device contained in one package. It is used to couple a signal between circuits without using electrical connections between the circuits.

Optics: Branch of physics dealing with light and vision.

Optoelectronic (Optronics) device: A device in the branch of electronics dealing with light which detects and/or is responsive to electromagnetic radiation (light) in the visible, infrared, and/or ultraviolet spectral regions; emits or modifies noncoherent electromagnetic radiation in these same regions; or utilizes such electromagnetic radiation for its internal operation.

Period: The length of time for one cycle of a repetitive frequency to occur.

Phase: Time relation between one wave and another or between a wave and a specific time.

Photoconduction: The difference between light current (I_L) and dark current (I_D) in a photodetector.

Photodiode: A diode which has been optimized for light sensitivity and is used to detect light.

Photoemission: Light releasing electrons from a surface.

Photometric brightness: See Luminescence.

Photon: Tiny pulses of light energy. An atom stimulated to an excited state emits photons of energy when it falls to the unexcited state. A quantum of light energy.

Phototransistor: A transistor (bipolar or field-effect) which has been optimized for light sensitivity and is used to detect light.

Photovoltaic: Generation of voltage with the use of dissimilar materials and light-sensitive material in response to radiation.

Plasma: Ionized gas.

Population inversion: Movement of an atom from a ground energy state to an excited energy state so that more atoms occupy the excited state than the ground state.

Power density: Irradiance.

Quantum efficiency (of a photosensitive device): The ratio of the number of carriers (electrons) generated to the number of photons incident upon the active region.

Radiance: The radiant flux per unit solid angle and per unit surface area normal to the direction considered. The surface may be that of a source, detector, or any other surface intersecting the flux.

Radiant energy: Energy generated by the frequency of radiation of electrons. The energy of light.

Radiant flux (θ_e): The time rate of flow of radiant energy.

Radiant power (ϕ): The time rate of flow of electromagnetic energy, measured in watts (W).

Radiation: The movement of energy in the form of rays or electromagnetic waves from the source.

Radiation pattern: The radiation pattern is a curve of the output radiation intensity plotted against the output angle.

Refraction: The bending of a light ray when the ray moves through the intersection of two different mediums such as air to water.

Source: The source of radiant energy, such as a light-emitting diode (LED).

Spatial coherence: Phase correlation of two different points across a wave front at a specific moment in time.

Spectral Output (of a Light-Emitting Diode): A description of the radiant-energy or light-emission characteristic versus wavelength.

Spectral Response (of a Photosensitive Device): A description of the electrical output characteristic versus wavelength of radiant energy incident upon the device.

Spontaneous emission: When an atom falls from an excited state to a ground state, and in doing so emits a photon of energy, the photon is said to have been spontaneously emitted.

Stimulated emission: When an external source of energy, such as a photon, causes an atom to move from an excited state to a lower state of energy and emit a second photon of energy.

Temporal coherence: Phase correlation of waves at a point in space at two instants of time.

Transfer ratio (of an Optically Coupled Isolator): The ratio of the dc output current to the dc input current.

Translucent: Material that allows some light to pass through while absorbing or reflecting the remainder.

Transparent: Material that allows almost all light to pass through.

Ultraviolet: Band of light wavelengths too short for response by a human eye.

Velocity: Rate of change of position in relation to time. Speed.

Visible light: Electromagnetic wavelengths, ranging from 380 nm to 770 nm, that can be seen by the human eye.

Visible-light-emitting diode (VLED): An optoelectronic device containing a semiconductor junction which emits visible light when forward-biased.

Wave: The motion of light as it travels through space.

Wavelength: The amount of space occupied by one cycle of an electromagnetic wave. Symbol (λ).

Index

AC Plasma Display: 5-22
Alphanumeric Display: 5-11
Angstrom: 1-5, 7
Anode: 3-11; 5-17, 19, 24
ASCII: 5-12
Atom: 2-10, 19; 3-7, 9
Avalanche Photodiode: 3-14; 7-19
Bar Code Reader: 1-19; 4-10; 8-17, 20
Beta: 7-15
Bias: 3-12; 7-2
Binary Code Indicators: 5-3
Bioluminescence: 2-11
Brightness Comparison: 2-9
Brushless Motors: 8-13
BVCBO: 7-15
BVCEO: 7-15
BVEBO: 7-15
Candela: 1-14; 6-1
Card Readers: 8-6
Cathode: 3-11; 5-17, 19, 24
Cathode Ray Tube (CRT): 5-1, 24
Cathodoluminescence: 2-12
Chemiluminescence: 2-11
Coherent Light: 2-18; 10-2
Color: 1-10; 8-2
Common Mode Rejection Ratio (CMRR): 4-20
Commutator: 8-13
Comparator: 6-16
Contrast Improvement Ratio: 6-10
Contrast Ratio: 6-10
Covalent Bond: 3-7
Current Transfer Ratio (CTR): 4-16
Cutoff Frequency: 4-19; 7-9
Cycle: 1-7
Decoder/Driver: 5-5
Depletion Region: 3-11, 14; 7-1
Detector: 1-9; 3-1
Diffraction Grating: 4-13
Displays: 5-2, 30
Distance Measurement: 8-4, 21; 10-13
Efficiency: 1-15; 2-16, 21; 3-12, 19; 7-3, 27
Electroluminescent Display: 5-21
Electromagnetic: 1-5; 2-9
Electron Gun: 5-24
Encoder: 8-10, 15, 17
Energy:
   Conversion: 1-17; 2-1; 3-1, 12; 7-26
   Gap: 3-9
   State: 2-10, 19; 10-3
   Transmission: 1-17
Excited State: 2-10, 20
Eye: 1-12; 4-5
Fall Time: 4-19; 7-9, 14
Fan-out: 4-17; 6-14
Fault Indicators: 5-2; 6-16
FET: 3-11; 7-17
Fiber Optic: 4-14; 9-3, 5, 19; 10-22
Filter: 4-9; 6-20, 21
Fluid Level Sensing: 8-10, 12
Fluorescent: 2-4, 12, 14
Flux: 4-14
Foot-candle: 1-14
Frequency: 1-7; 4-19
   Response: 4-18; 7-11
   Spectrum: 1-5
Gas Discharge:
   Display: 2-5; 5-17
   Lamp: 2-5, 14, 15, 16
Geodimeter: 10-15
Grid, Control: 5-19, 25, 27
Ground Loop: 4-21
Guidance System: 8-25, 26
Headlight Dimmer: 7-24
Hertz (Kilo, Mega, Giga): 1-7
Hole: 3-9
Hologram: 10-26
Ignition System: 8-14, 16
Illumination: 1-16
Incandescent: 2-4, 12
Incoherent Light: 10-1
Index of Refraction: 9-4
Infrared: 1-5, 6-19; 8-22; 9-11
   Remote Control: 9-7
Injection Luminescence: 2-16
Ink Jet Printer: 5-28
Insertion Loss: 4-15
Intensity: 1-2, 12, 15; 2-21
Interruptible System: 1-18; 4-2
Ionize: 2-14; 5-17, 22
Isolation: 4-20, 23
Junction Capacitance: 3-11, 12; 7-1, 8, 13
Laser: 1-6; 2-8, 18; 10-1
LED: 2-7, 16; 6-1
Lens: 2-21; 3-3, 19; 4-9; 7-5
Light: 1-4, 5, 8; 2-1, 9, 17
Liquid Crystal (LCD): 2-8, 18; 5-14
Logic Level Indicators: 5-3; 6-13
Lumen: 1-14

INDEX

Luminescence: 2-11
Luminous Intensity: 1-14; 6-1
Lux: 1-14
Maser: 2-18; 10-10
Medium: 1-8; 3-1, 2; 4-1, 5
Mercury Vapor: 2-16; 5-17
Micron: 1-5
Microwave: 1-5
Mirrors: 4-11; 10-4
Monochromatic: 2-18; 10-1
Multiplexing:
    Display: 5-5
    Time Division: 9-22, 23
N-type Material: 3-9
Neon: 2-3, 5, 15; 5-17
Noise: 4-22
Noise Equivalent Power: 4-23
Non-interruptible System: 1-18; 4-3
Operational Amplifier: 6-5, 20; 9-15
Optical Filters: 4-9
Optically Coupled System: 4-1
Opto-isolator (Opto-coupler): 4-6, 7, 22; 9-2, 15
P-type Material: 3-9
Peltier emf: 3-4
Period: 1-7
Phase Locked Loop (PLL): 9-11
Phosphorescence: 2-12
Photo:
    coupler: 4-7
    cathode: 3-13
    current: 7-2, 3
    diode: 3-11; 7-1
    emissive: 3-4, 13
    field-effect transistor: 3-11; 7-17
    metric: 1-12; 2-8, 21
    multiplier: 3-13
    resistive: 3-4, 9, 11
    thyristor (SCR): 3-11; 7-21
    transistor: 3-11; 7-13
    voltaic: 3-4, 11
Photon: 1-6; 2-19; 3-11, 13; 10-2
PIN Diode: 7-1
PN Junction: 2-16; 3-11; 7-1
Population Inversion: 2-20; 10-4
Prisms: 4-12
Pumping: 2-20; 10-3
Quantum Detector: 3-4
Quantum Efficiency: 2-16; 3-13; 7-3
Radiant Energy: 1-4
Radiation: 1-5; 2-1, 11
Radiometric: 1-12, 15
Range Finding System: 8-22
Reference Diode: 7-19

Reflective Light Sources: 2-2, 18
Reflectivity: 8-2
Reflector: 2-21; 4-7
Reliability: 5-17, 30; 6-6
Resistivity: 3-9, 12
Retina: 1-14
Retroreflector: 10-13
Rise Time: 4-19, 7-9, 14
Second (Milli, Micro, Nano, Pico): 1-7
Secondary Emission: 3-13
Seebeck emf: 3-4
Semiconductor Materials: 2-10, 17; 3-7; 5-13
Sensitivity: 7-15
Seven-segment Display: 5-4, 22
Shell Structure of Atoms: 2-10; 3-7
Signal-to-Noise Ratio: 4-23
Solar Cell: 1-17; 3-12; 7-26
Source, Light: 2-1, 17
Spatial Coherence: 10-1
Spectral Distribution: 2-20; 3-17; 4-3; 7-3
Spectral Response: 3-17; 4-3; 7-3
Spectrum: 1-5, 10
Speed: 1-8, 3-12; 8-1
Spontaneous Emission: 2-19; 10-2
SRCBO: 7-15
SRCEO: 7-15
Stimulated Emission: 2-18, 19; 10-3
Strobe: 5-9
Switching Time: 4-19; 5-29
Temperature: 2-13; 3-10; 6-6
Temporal Coherence: 10-1
Thermal Detector: 3-4
Thermocouple: 3-4
Thermopile: 3-6
Thomson emf: 3-4
Transmission Medium: 1-2, 18; 4-1, 5; 8-1; 9-1, 3
Transmissivity: 8-4
TTL: 4-17
Tungsten Filament: 2-9, 12; 3-17; 5-19
Twilight Detector: 7-25
Ultraviolet: 1-5; 2-14
Vacuum Fluorescent Display: 5-19
Valence Electrons: 2-11; 3-7
VCE(sat): 7-15
Viewing Angle: 2-21; 3-16, 18; 5-28; 7-5, 6
Visible Spectrum: 1-10
VLED: 2-7; 5-3; 6-1, 7
Voltage Comparator: 6-16; 8-8, 12
Voltage Controlled Oscillator (VCO): 9-11
Watt: 1-15
Wavelength: 1-6